FLORA OF TROPICAL EAST AFRICA

DRYOPTERIDACEAE

J.P. Roux[1], Monika Shaffer-Fehre[2] & Bernard Verdcourt[3]

Terrestrial, epilithic or low-level epiphytic plants with dictyostelic, erect or creeping, often stoloniferous rhizomes bearing scales. Fronds spaced or tufted, monomorphic or dimorphic; stipes mostly dark at the base, adaxially sulcate or with a central ridge in *Lastreopsis*, with two larger circular vascular bundles dorsally and two or more smaller ones ventrally, not articulated, variously set with scales and/or hairs; rachis sulci open or, if present, closed to the sulci of the lower order axes; lamina 1–4-pinnate, anadromous and/or catadromous, the basal pinnae often basiscopically developed, often with proliferous buds along the lamina axes, variously set with scales, hairs and glands along the veins and lamina surfaces; venation forked or pinnately branched, free or anastomosing, with or without included veinlets, veins mostly ending near the lamina margin. Sori circular or elongate, dorsally on veins or at vein endings, often on an abbreviated vein branch, receptacle with or without simple paraphyses; indusium reniform, circular, elongate, peltate or marginally attached, or exindusiate.

As treated here the family consists of 28 genera and approximately 1050 species.

The Dryopteridaceae as construed here is often treated as two independent families, Dryopteridaceae *sensu stricto* consisting of *Didymochlaena, Cyrtomium, Polystichum, Nothoperanema, Dryopteris* and *Arachniodes*, and Tectariaceae, which includes *Tectaria, Triplophyllum, Lastreopsis, Ctenitis, Hypodematium* and *Megalastrum*. Tectariaceae is characterised by raised costae and costules, presence of acicular and/or ctenitoid hairs, and a chromosome number based on n = 40 or 41. In Dryopteridaceae *sensu stricto* the costae and costules are adaxially sulcate, never bear acicular or ctenitoid hairs, and have a chromosome number based on n = 41.

1. Lamina axes and veins variously set with hairs and scales, at least some of the hairs acicular (needle-shaped, stiff and pointed) 2
   Lamina axes and veins variously set with hairs and scales, the hairs never acicular 7
2. Pinna-rachises flat or convex 3
   Pinna-rachises sulcate, often with a central ridge 4
3. Rhizome short, erect to suberect; fronds tufted; venation reticulate .... 1. **Tectaria** (p. 2)
   Rhizome slender, creeping; fronds spaced; venation free (rarely with casual anastomoses near margin) .... 2. **Triplophyllum** (p. 8)
4. Lamina axes sulci centrally ridged .... 3. **Lastreopsis** (p. 12)
   Lamina axes sulci not centrally ridged 5
5. Lamina axes and veins variously set with narrow scales and ctenitoid hairs (see Fig. 5.9, p. 16) .. 4. **Ctenitis** (p. 15)
   Lamina axes and veins variously set with acicular and glandular hairs 6

[1] Compton Herbarium, Private Bag X7, Claremont 7735, South Africa. *Polystichum, Dryopteris*
[2] c/o Royal Botanic Gardens Kew, UK. *Tectaria, Triplophyllum, Lastreopsis, Ctenitis, Megalastrum, Cyrtomium, Nothoperanema*
[3] c/o Royal Botanic Gardens Kew, UK. *Hypodematium, Didymochlaena, Arachniodes*

6. Lamina up to 35 cm long; acicular hairs unicellular; glandular hairs absent .... 5. **Hypodematium** (p. 17)
   Lamina up to 120 cm long; acicular hairs pluricellular; glandular hairs present .... 6. **Megalastrum** (p. 20)
7. Pinnae and pinnules articulated; sori elliptic .... 7. **Didymochlaena** (p. 22)
   Pinnae and pinnules not articulated; sori orbicular 8
8. Indusium peltate, if exindusiate the lamina somewhat acroscopically developed 9
   Indusium reniform, if exindusiate the lamina then basiscopically developed 10
9. Lamina 1-pinnate; venation reticulate .... 8. **Cyrtomium** (p. 24)
   Lamina mostly 2- or more pinnate; venation free 9. **Polystichum** (p. 26)
10. Lamina hairs (trichomes) multiseptate at base ... 10. **Nothoperanema** (p. 33)
    Lamina hairs (if present) uniseptate at base 11
11. Pinnule margins crenate to serrate .... 11. **Dryopteris** (p. 35)
    Pinnule margins aristate-dentate .... 12. **Arachniodes** (p. 49)

1. TECTARIA

Cav. in Ann. Hist. Nat. Madrid 1: 115 (1799); Holttum in K.B. 38: 108, t. 1 (1983) & in Flora Malesiana ser. 2, 2(1): 39–100 (1991)

Medium to large terrestrial ferns; rhizomes woody, erect to suberect, with firm linear to lanceolate scales. Fronds spaced or tufted, dimorphic; fertile fronds of some species taller than sterile, lamina of blade contracted; blade lanceolate to deltoid-pentagonal, simple to 1-pinnate or 2–4 pinnatifid, basal pair of pinnae often developed basiscopically; divisions usually broad with entire or sparingly incised margins, ultimate segments crenate but not aristate; veins anastomosing, areoles with or without free included veinlets. Sori mostly round, arranged in open rows, median, or scattered; indusium peltate, reniform or sometimes apparently lacking; sporangia with an annulus of ± 14 cells. Spores monolete, oval, perispore echinate or tuberculate; indusium peltate, reniform or absent.

Over 200 species tropical in distribution adapted to moist and windless habitats.

The general process of phylogenetic differentiation has been a simplification of the frond and a correlated anastomosis of the veins (Copeland, Philipp. J. Sci. 2C.: 410 (1907)).
A finding during the present work was that measuring the lengths of sporangial stalks and number/lengths of trichomes/paraphyses below the capsule affords such a good taxonomic character for separating the taxa, that even their absence from *T. angelicifolia* is diagnostic.

1. Gemmae or their scars on adaxial rachis, mostly on junction with pinnae and often on pinna-costa ...... 2. *T. gemmifera*
   Gemmae absent 2
2. Fertile blade with 6–8 pairs of pinnae .... 1. *T. coadunata*
   Fronds with 5 or fewer free pinnae 3
3. Pinnae of 1 opposite pair; fern of low altitudes on coast (< 150 m) .... 4. *T. puberula*
   Pinnae of 1–5 pairs; ferns of higher altitudes (> 400 m) 4
4. Ultimate segments rounded .... 3. *T. angelicifolia*
   Ultimate segments acute .... 5. *T. torrisiana*

1. **Tectaria coadunata** (*J.Sm.*) *C.Chr.* in Contr. U.S. Nat. Herb. 26: 331 (1931); Fraser-Jenkins, New species Syndrome Indian Pter.: 242 (1997). Type: Nepal, *Wallich* 377 (K-Wall.!)

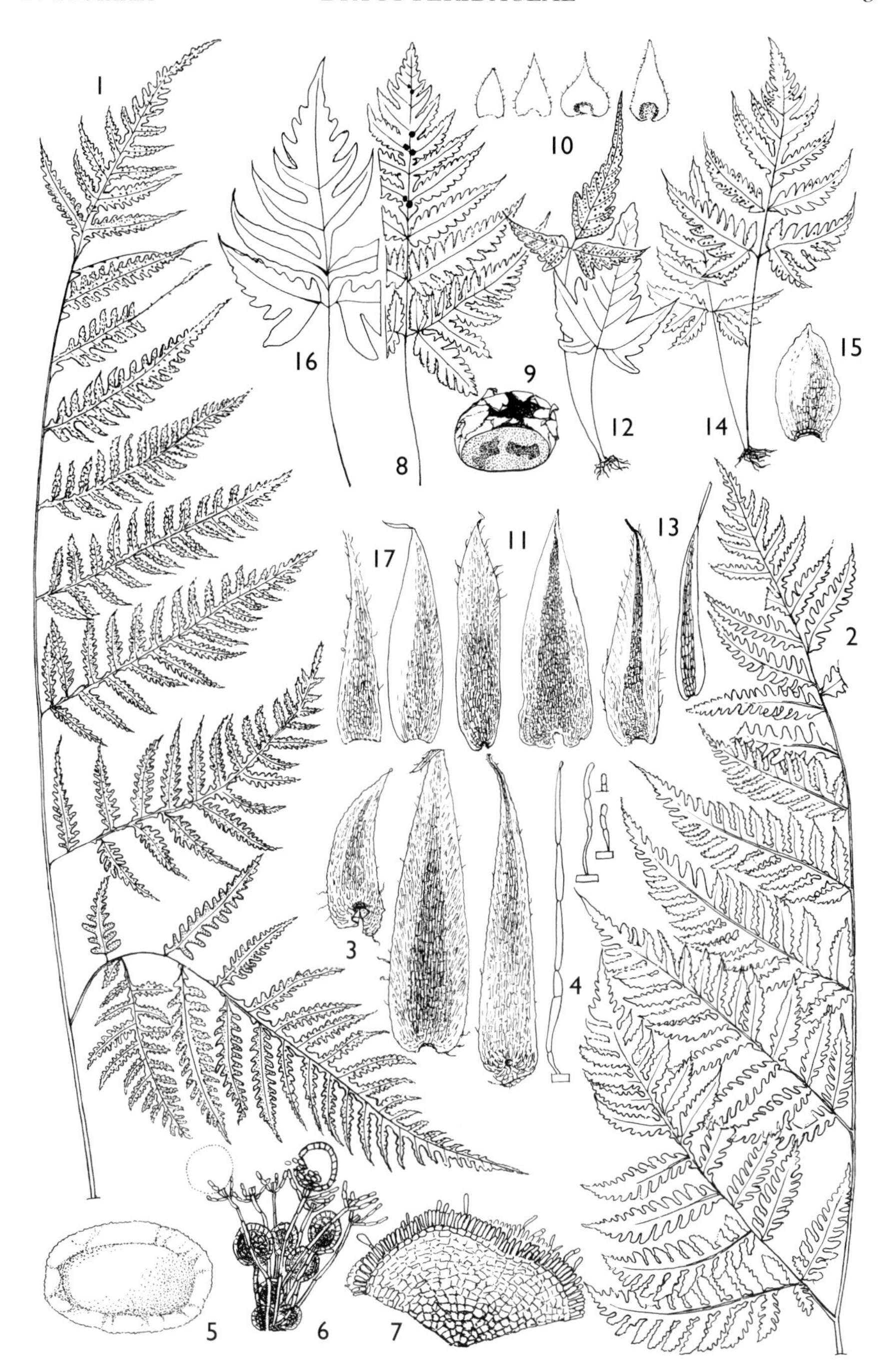

FIG. 1. *TECTARIA COADUNATA* — **1**, fertile frond, × $^1/_6$; **2**, sterile frond, × $^1/_6$; **3**, stipe scales, × 5; **4**, catenate and glandular hairs, × 85; **5**, spore, × 900 ; **6**, part of sorus, × 15; **7**, indusium, × 14. *TECTARIA GEMMIFERA* — **8**, fertile frond, × $^1/_6$; **9**, gemma, × 3; **10**, gemma scales, × 5; **11**, stipe scales, × 5. *TECTARIA PUBERULA* — **12**, fertile and sterile frond, × $^1/_8$; **13**, stipe scale, × 5. *TECTARIA ANGELICIFOLIA* — **14**, fertile and sterile frond, × $^1/_8$; **15**, stipe scale, × 5. *TECTARIA TORRISIANA* — **16**, fertile and sterile frond, × $^1/_8$; **17**, stipe scales, × 5. 1–7 from *van Someren* 20; 8–11 from *Peter* 10266; 12–13 from *Faden et al.* 70/808; 14–15 from *de Boer et al.* 788; 16–17 from *Faden et al.* 70/366. Drawn by Monika Shaffer-Fehre.

Terrestrial; rhizome erect, to 20 cm high, covered in scales (not seen in East African material). Fronds tufted, dimorphic, to 1.5 m long; stipe straw-coloured, later mahogany-coloured, to 70 cm long, near base with dark brown ovate scales 1–10 × 0.4–2 mm, acuminate, with paler margins, stipe near apex with multicellular glandular trichomes ± 0.3 mm long; lamina mid- to dark green, subcoriaceous, sterile lamina deltoid in outline, to 68 × 64 cm, the fertile lamina almost lanceolate above the wide basal pinnae, 50–108 × 43–84 cm, 3-pinnatisect; pinnae in 6–8 pairs, the lowermost to 42 × 27 cm and longest, with stalks of 1.5–3 cm, others much shorter and gradually adnate to rachis by basiscopic half; basal pinnules largest, to 16 cm, with basiscopic part of pinna broader, basal few pinnules free, other pinnules joined by a continuous wing; ± straight free veinlets in central areoles, absent along costae; lamina with some hairs near margin plus glandular trichomes to 0.4 mm restricted to in and near sinus, and each areole with a single trichome to 0.3 mm; sori almost marginal, 1–2 mm in diameter; indusium round, persistent, sometimes flabellate. Fig. 1: 1–7 & Fig. 2: 1.

UGANDA. Bunyoro District: Budongo Forest, Sep. 1992, *Sheil* 1349! & Budongo Forest reserve close to Sonso R., Sep. 1995, *Poulsen et al.* 964!; Kigezi District: Bwindi National Park, Ishasha Gorge, May 1995, *Poulsen et al.* 819!

KENYA. Meru District: upper Meru forest, without date, *van Someren* 680!; Machakos/Masai District: Chyulu Hills, May 1938, *van Someren* 19! & 52!

TANZANIA. Morogoro District: N Uluguru Mts, Mkungwe Hill, July 1970, *Faden et al.* 70/365!

DISTR. **U** 2; **K** 4, 6; **T** 6; SW China, NE India, Thailand, Malaysia

HAB. Moist forest; 1000–1900 m

USES. None recorded

CONSERVATION NOTES. Widespread; least concern (LC)

SYN. *Aspidium coadunatum* Hook. & Grev., Icon. Filic.: 202 (1831), *non* Kaulf. 1824, *nom. illeg.*
Sagenia coadunata J.Sm. in J. Bot. (Hooker) 4: 184 (1841), *nom. nov.*
Sagenia macrodonta Fée, Gen. Fil.: 313, t. 24 (1852). Type as for *Tectaria coadunata*
Tectaria macrodonta (Fée) C.Chr., Ind. Suppl. 3: 181 (1934)

NOTE. Blunt-rounded lobes of the decurrent lamina fill gaps between tissues of blade before the lamina expands; probably an adaptation to support increased photosynthesis in the deep shade of the plant's habitat.

2. **Tectaria gemmifera** (*Fée*) *Alston* in J. Bot. 77: 288 (1939); Schelpe, F.Z., Pter.: 234, t. 64D (1970); Burrows, S. Afr. Ferns: 324, fig. 54.2, fig. 78, map (1990); Faden in U.K.W.F. ed. 2: 35 (1994); J.P. Roux, Conspect. southern Afr. Pterid.: 132 (2001). Type: Madagascar, 'habitat in insula Madagascariens', *Pervillé* s.n. (missing)

Terrestrial; rhizome to 13 cm high, 2 cm in diameter, with dark brown lanceolate scales 5–10 × 1 mm, with pale borders. Fronds in tufts of 4–9, to 1.8 m long; stipe pale brown, 70–72 cm long, 5 mm in diameter at base, adaxially grooved, basal 3–5 cm covered in scales similar to those of rhizome; lamina membranous to coriaceous, dark green, deltoid-pentagonal, 45–100 × 30–80 cm, 3-pinnatifid; pinnae in 4–6 pairs, basal pinna pair the longest and stalked, terminal segment pinnatifid; pinnules to 18 cm long, the basal the largest, divided in up to 10 secondary pinnules; ultimate segments oblong, falcate, with crenate margins; veins and veinlets anastomosing; free included veinlets in areoles not frequent, absent along the rachis and rare in the triangular areoles between rachis and costa; small glandular hairs ± 0.5 mm long thinly scattered on axes from rachis to costa, denser at pinna base and pinna margins where it is decurrent between pinna segments, on sinus margin and adaxially, above sinus, on rachis, pinna costa and costa, resembling velvet; minute scales sometimes present in central areoles; glabrous when mature. Gemmae sometimes present at pinnae bases or along pinnae costa. Sori inframarginal to median, in up to 4 pairs per segment, 0.5–2(–3) mm in diameter; indusium small, reniform, minutely ciliate. Spores pale brown, monolete, perispore winged, cristate. Fig. 1: 8–11 & Fig. 2: 5.

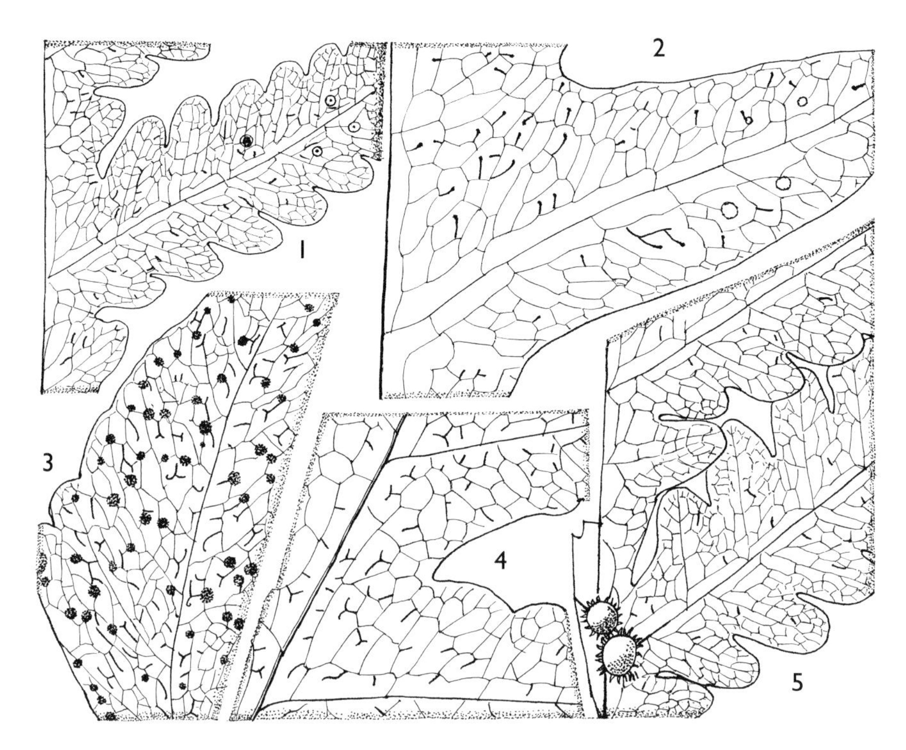

FIG. 2. *TECTARIA*: differently shaped free veinlets; all × 1.2. **1**, *T. coadunata* ; **2**, *T. torrisiana* ; **3**, *T. puberula*; **4**, *T. angelicifolia*; **5**, *T. gemmifera*. 1 from *van Someren* s.n.; 2 from *Faden et al.* 70/366; 3 from *Faden et al.* 70/808; 4 from *Sheil* 1750; 5 from *Peter* 10266. Drawn by Monika Shaffer-Fehre.

UGANDA. Karamoja District: Mt Kadam, Apr. 1959, *J. Wilson* 775!; Bunyoro District: Budongo Forest, Apr. 1994, *Sheil* 1807!; Mbale District: Kami [Khami] R. valley, Apr. 1951, *Wood* 190!

KENYA. Trans Nzoia District: Cherangani Hills, Kabolet, Aug. 1963, *Tweedie* 2702!; Nyeri District: Gaturu, Aug. 1963, *Mathenge* 228!; Masai District: Chyulu Hills, main forest N, Dec. 1993, *Luke & Luke* 3883!

TANZANIA. Lushoto District: E Usambara Mts, Amani-Sigi Forest Resereve, along Sigi R., Oct. 1986, *Borhidi et al.* 86/308!; Kigoma District: Gombe National Park, Kakombe Valley, Dec. 1963, *Pirozynski* P104!; Iringa District: Mwanihana Forest Reserve above Sanje, Sep. 1984, *D.W. Thomas* 3685!

DISTR. **U** 1–4; **K** 1, 3–7; **T** 2–4, 6–8; Congo-Kinshasa, Rwanda, Burundi, Angola and south to South Africa; Madagascar, ?southeast Asia

HAB. Moist forest, especially near streams, also in gallery forest; terrestrial, occasionally in rock crevices; may be locally common; 600–2550 m

USES. None recorded

CONSERVATION NOTES. Widespread; least concern (LC)

SYN. *Sagenia gemmifera* Fée, Mem. Foug. 5: 313 (1852)
Aspidium coadunatum Kaulf. var. *gemmiferum* (Feé) Kuhn, Filic. Afr.: 128 (1868)
Aspidium gemmiferum (Feé) Ching in Bull. Fan. Mem. Inst. Biol. 10: 237 (1941)

NOTE. Sori of fertile specimens of *T. gemmifera* vary from barely 0.5 mm diameter (gemmae present) to almost 3 mm diameter (gemmae absent); the precociously fertile or, indeed, sterile specimens will almost always be covered by the largest (7 mm in diameter) or the most numerous (to 25 in *Pirozynski* P104) gemmae; i.e. a large physiological debt arises to account for the production of gemmae; but only one 5 mm gemma and a fully sporiferous blade in *Thomas* 1498. In the presence of gemmae sori will occasionally develop only in segments near tips of pinnules. From the point of reproduction, large sori in the absence of gemmae or many gemmae on sterile or precociously fertile fronds, may offer an equal chance of creating progeny. An increased dissection of the blade can carry a larger number of gemmae.

A caution must be sounded to check gemmae closely. Due to occasional infestation by psychids (Psychideae, insects) specimens have been wrongly identified as *T. gemmifera*. Psychids coat their coccoon with "locally available" materials. On fern blades spores serve this purpose. As the larva clings to the vascular supply, a thick coat of spores is sometimes mistaken for a gemma.

In the literature 'white hairs' are mentioned with the indumentum; during this survey they were observed just once (*Pirozynski* P104), a fine down of white, septate hairs (lost early) arranged regularly, following the vascular system and admixed with stout glandular cells.

3. **Tectaria angelicifolia** (*Schumach.*) *Copel.* in Phil. Journ. Sci. 2: 410 (1907); Alston, Ferns W.T.A.: 74 (1959); Tardieu in Fl. Cameroun 3: 289 (1964); J.P. Roux, Conspect. southern Afr. Pterid.: 132 (2001). Type: Guinea, sine coll. 304 (C, holo.)

Terrestrial; rhizome erect or shortly creeping, with brown lanceolate scales 3–5 × 0.5–0.8 mm, apex attenuate, margin sparsely denticulate. Fronds spaced at short intervals or tufted; stipe stramineous, 20–30 cm long, with few brown scales 5–6 × 0.5–0.8 mm at base, a few more scales in adaxial groove into rachis, and higher up when young with dense short, gland-tipped hairs, rapidly glabrescent; lamina herbaceous to coriaceous, triangular, 20–32 × 20–28 cm, 3-pinnatifid; pinnae in 1–5 pairs, the lowermost opposite, stalked for up to 5 cm, falcate, 15–18 × 7–8 cm; terminal pinna deeply lobed at its base, the lobes gradually decreasing in size; basiscopic pinnule of lowermost pinna more developed than acroscopic one, to 12 cm long and almost free, other basiscopic pinnules gradually decreasing; ultimate segments broad, shallow, rounded; free veinlets potentially present in all areoles of lamina, including areoles parallel and near middle of rachis and costa; veinlets variable from short bent to long curved to T-shaped and minutely branching; with septate, glandular hairs to 0.5 mm long along margin of lamina and on vascular axes. Sori median in most pinnules of lowest pinna, round, 1.2–1.5 mm in diameter, smaller in more distal parts of lamina; indusia often fugitive. Fig. 1: 14–15 & Fig. 2: 4.

UGANDA. Acholi/Bunyoro District: Murchison Falls National Park, May 1993, *Sheil* 1750!; Toro District: Mwamba Forest, Dec. 1957, *Allen* 3691!; Mengo District: Mabira Forest Reserve, Nov. 1938, *Loveridge* 69!

TANZANIA. Lushoto District: Usambara, Silai, Feb. 1896, *Holst* 2306!; Ulanga District: Ifakara, Matundu Forest Reserve, Sep. 1994, *Kisena* 1454!

DISTR. **U** 1, 2, 4; **T** 3, 6; Sierra Leone to Sudan and south to Angola and Congo-Kinshasa

HAB. Moist forest; 400–1200 m

USES. None recorded

CONSERVATION NOTES. Widespread; least concern (LC)

SYN. *Polypodium angelicifolium* Schumach. in Vid. Selsk. Afh. 4: 228 (1827), as *angelicaefolium*
Aspidium angelicifolium (Schumach.) C.Chr., Ind.: 64 (1905)

4. **Tectaria puberula** (*Desv.*) *C.Chr.* in Dansk Bot. Ark. 7: 67 (1932). Type: Réunion, 'habitat in insula Borboniae', no collector indicated, 356 (P, holo.)

Rhizome long-creeping, up to 7 mm in diameter, covered in dark brown scales up to 0.7 mm at base with attenuate apex. Fronds spaced; stipe 23 cm long in fertile, 14 cm long in sterile fronds, with scales 2.5–4 × 0.2 mm near base; lamina tripartite-palmate; fertile: basal pinna pair together 16 × 3.2 cm, pinnae basiscopically produced, stalked for ± 1 mm, terminal pinna 14.5 × 4.8 cm with lobed margin; sterile lamina more foliose, all elements fused, also basiscopicaly produced and up to 17 × 8 cm, basal pinnules together 17 × 5 cm, sessile; rachis brown with adaxial groove containing minute dark brown glands; veins anastomosing, free veinlets in all areoles even in those parallel to rachis; adaxially and abaxially with sparse stiff hairs to 0.3 mm long, sometimes with small scales. Sori median on fused pinnules and on smaller units, never on tip of free veinlet, but rather on branch points of veins, brown, ± 1 mm in diameter, occasionally confluent. Fig. 1: 12–13 & Fig. 2: 3.

KENYA. Kwale District: Shimba Hills National Park, Sheldrick's Falls, Apr. 1968, *Magogo & Glover* 620! & idem, Nov. 1970, *Faden et al.* 70/808!
TANZANIA. Pemba, Semewani, Aug. 1929, *Vaughan* 451!
DISTR. **K** 7; **P**; Madagascar, Mascarene Islands
HAB. Spray zone by falls in moist forest (one record); 1–150 m
USES. None recorded
CONSERVATION NOTES. Not evaluated (NE)

SYN. *Polypodium triphyllum* Desv. in Mag. Nat. Berlin 1811: 315 (1811) & in Journ. Bot. Appl. (Paris) 4: 260 (1824), *non* Jacq. 1788
Aspidium puberulum Desv. in Mem. Soc. Linn. Paris 6: 245 (1827); C. Chr., Ind.: 89 (1905), *non* Gaud. 1827 *nec* Fée 1865 ; *nomen novum* for *Polypodium triphyllum* Desv.

5. **Tectaria torrisiana** *Shaffer-Fehre* **sp. nov.** planta matura quam *T. lawrenciana* $^1/_2$–$^1/_3$ minora. Lamina fertilis paribus duobus pinnarum liberis stipitatis (nec pare uno tantum) ornata, pinnula infima basin versus producta atque ad apicem anguste cordata (nec rotundata) pinnulis proximis similis. Venae atrorubrae; areolae juxta rachidem costamque pinnarum locatae angustae atque ad eas parallelae nec carentae neque inordinatae. In medio laminarum venulae liberae plerumque leniter curvatae, raro rectae, rarissime tantum 'T' formantes. Type: Tanzania, Morogoro District: N Uluguru Mts, Mkungwe Hill, *Faden* 70/366 (K!, holo.)

Terrestrial; rhizome erect or shortly creeping, woody, to 1 cm in diameter, with dense lanceolate scales 5–8 × 0.7–1.7 mm, attenuate, either mid-brown and uniform or with darker centre and thin paler margin. Fronds tufted, somewhat fleshy, to 120 cm long; stipe reddish brown, lowest 4 cm almost black, 27–55 cm long, shorter (< 20 cm) in sterile fronds, mostly glabrous but with a few scales near base, similar to those of rhizome; lamina ovate in outline, 32–50 × 24–34 cm, 2-pinnate to 3-pinnatifid, lustrous; pinnae 2 free pairs on each side below pinnatifid apex, to 27 × 23 cm; pinnules sharply upturned, acuminate, basiscopic pinnule larger (to much larger), terminal pinna deeply lobed-pinnate and to 30 × 22 cm; ultimate segments to 2.5 × 2 cm, falcate and acute; free veinlets simple or occasionally forked as a 'T' with the tip often at a slight angle, thickened as for hydathode; lamina glabrous except for septate glandular hairs in grooves of costa and on costules and margins in sinus. Sori median, mostly near tips of pinnules, on end of free veinlets and on branching points, 1–1.8 mm in diameter; indusium reniform. Fig. 1: 16–17 & Fig. 2: 2.

TANZANIA. Morogoro District: N Uluguru Mts, Mkungwe Hill, July 1970, *Faden et al.* 70/366! & Uluguru Mts, Mwere valley, Sep. 1970, *Faden et al.* 70/598! & Uluguru Mts, Mlulu Valley, Mar. 1972, *Pócs & Mwanjabe* 6558/D!
DISTR. **T** 6; only known from the Uluguru Mts
HAB. Moist forest; 1050–2000 m
USES. None recorded
CONSERVATION NOTES. Of limited distribution in a rapidly disappearing habitat: vulnerable (VU-D2)

NOTE. Named in honour of Maria Torris, 'Loho', actress, reader and green-fingered godmother who raised the orphaned, teenage author with great devotion. [1]

Faden thought this might be "Tectaria sp. nov. aff. *T. lawrenceana* (Moore) C.Chr.". That species has pinnae that are ± blunt-tipped and do not have distinct areoles parallel to the costa; the free veinlets here had a "T"-shape throughout. The stalk of the "T" was often shorter and the cross bar longer than in *T. angelicifolia*, with the arms upturned. *T. torrisiana* is distinct due to most of its pinnae /pinnules having entire margins with upward-turned acute tips.

[1] With thanks to Melanie Thomas for the Latin translation.

2. TRIPLOPHYLLUM

Holttum in K.B. 41(2): 239 (1986)

Dryopteris subgen. *Ctenitis*, group of *D. protensa* (Sw.) C.Chr., Monogr. *Dryopteris*, pt. II: 91 (1920)

Ctenitis sensu Tard. in Mém. I.F.A.N. 28: 129 (1953) pro parte; *sensu* Alston, Ferns W.T.A.: 70 (1959) pro parte

Rhizome long-creeping, slender, with narrow *Tectaria*-like scales, not clathrate. Fronds spaced; vascular structure of rhizome and stipes as in *Tectaria*. Lamina of mature plants either deltoid-pentagonal or somewhat elongate; in almost all cases basal pinnae fully pinnate, their basal basiscopic pinnules in some species bearing leaflets of the fourth order; ultimate leaflets almost always asymmetric at their bases, basiscopic base narrowly cuneate; venation in most species free, in a few species anastomosing but not forming regular, narrow costal areoles, free veins absent from areoles; cylindric glands, of the kind universal in *Ctenitis*, absent. Sori usually round; indusia reniform, in some species very small or absent; spores bearing thin, translucent wings; base chromosome number n=41.

20 species in the tropics of Africa and America.

1. Fronds of mature plants tripartite or broadly deltoid-pentagonal, the basal pinnae much longer than the next pair 2
   Fronds of mature plants ± elongate, basal pinnae not much longer than the pair next above them .... 3. *T. fraternum*
2. Stipe 50–100 cm long; lamina 47–60 × 60–74 cm .... 2. *T. troupinii*
   Stipe 18–55 cm long; lamina 18–36 × 15–39 cm 3
3. Basal pinnae widely over-reaching (handle bar) the remainder of blade above, strongly curved upwards .... 1. *T. protensum*
   Basal pinnae wider than remainder of blade, gently curved upward .... 4. *T. vogelii*

1. **Triplophyllum protensum** (*Sw.*) *Holttum* in K.B. 41(2): 247 (1986). Type: Sierra Leone, *Afzelius* s.n. (S, holo.)

Terrestrial; rhizome long-creeping, 3–5 mm in diameter, with dense overlapping gland-tipped broadly triangular scales 1–2 × 1 mm. Fronds spaced 0.5–3.5 cm apart, erect; stipe brown, 18–55 cm long, base with lanceolate scales 4–5 × 0.4–1 mm, upper part with dense fine ctenitoid / catenate hairs 1–2 mm long, glabrescent; lamina dark green above, yellowish green below, thin to sub-coriaceous, deltoid-pentagonal in outline, with wide basal pinna pair strongly curving upwards, 18–36 × 15–32 cm, 3–4-pinnatifid; rachis not winged, grooved almost to apex; pinnae up to ± 12 on each side of the rachis, apex gradually decrescent, pinnae closely aligned, not overlapping; basal

FIG. 3. *TRIPLOPHYLLUM PROTENSUM* — **1**, habit, also showing second pinna, × $^1/_4$; **2**, basal acroscopic pinnules of second pinna, × 1; **3**, margin of indusium, × 50; **4**, gland-tipped scale, × 9; **5**, catenate hair with glandular tip, × 19. *TRIPLOPHYLLUM TROUPINII* — **6**, basal pinna of frond, × $^1/_4$; **7**, second pinna, × $^1/_4$; **8**, basal acroscopic pinnules of second pinna, × 1; **9**, stipe hairs, × 5; **10**, trichomes from costules, × 5. *TRIPLOPHYLLUM FRATERNUM* — **11**, second pinna, × $^1/_4$; **12**, basal acroscopic pinnule of second pinna, × 1; **13**, marginal and underside trichomes, × 200; **14**, stipe scale, × 12. *TRIPLOPHYLLUM VOGELII* — **15**, second pinna, × $^1/_4$; **16**, basal acroscopic pinnules of second pinna, × 1; **17**, indusium; **18**, stipe scale, × 19. 1–5 from *Loveridge* 186; 6–10 from *Thomas* 1347, 1348; 11–14 from *Faden* 70/273; 15–18 from *Thomas* 903, 1345. Drawn by Monika Shaffer-Fehre.

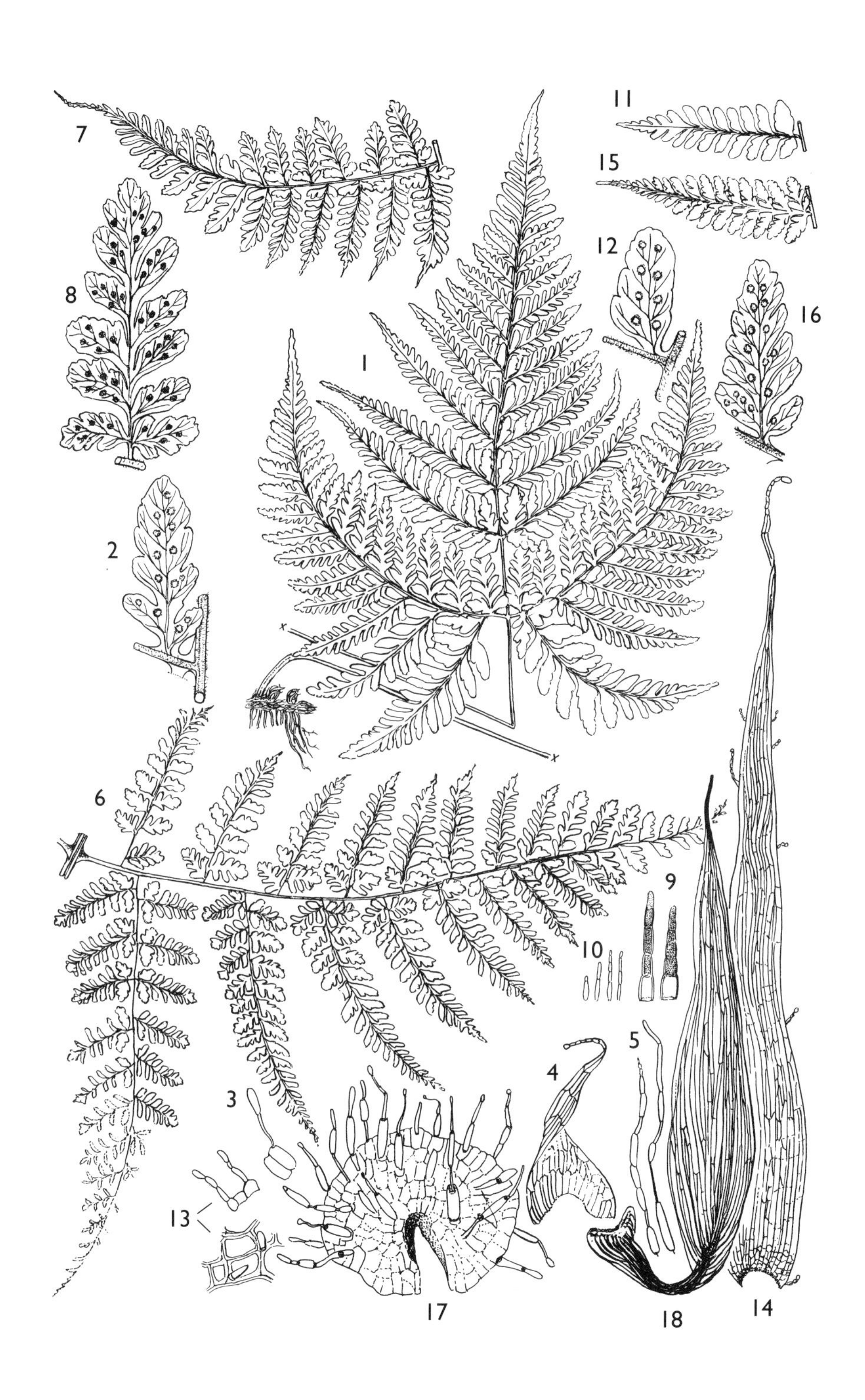

7
11
15
12
8
16
1
2
6
9
10
5
4
3
13
17
18
14

pinna to 26 × 13 cm, with well-developed basiscopic pinnule; second pair shorter and less pinnatifid, central pinnae to 12 × 4 cm, 1-pinnate except for the basal pinnules which are 2-pinnatifid; ultimate segments of central pinnae slightly falcate, 7 mm wide, crenate to entire, apex obtuse; margin of lamina framed by single line of achlorophyllous cells; veins and veinlets forked, not reaching margin; indument on abaxial surface sparse, denser on venation, of ctenitoid hairs 0.4–1.3 mm long, on adaxial surface with only a few trichomes 1–2 mm long near forks of veins. Sori submedian, near tip of acroscopic lateral veinlet, 1–1.5 mm in diameter; indusium kidney-shaped with lateral notch, 0.8 mm in diameter, from notch lines of square cells fan out, almost every second cell tipped by glandular hair ± 4 cells long; spores monolete, oval, 0.04 mm long, exospore with crest of few, blunt spines. Fig. 3: 1–5.

UGANDA. Bunyoro District: Budongo Forest, Dec. 1938, *Loveridge* 186!; Mengo District: 1.5 km NE of Nansagazi, Sep. 1969, *Faden et al.* 69/1028!; Masaka District: Sese, Bugala Island below Mkalive spring, Sep. 1997, *Lye & Katende* 22711!
DISTR. **U** 2, 4; widely distributed in West Africa; Madagascar
HAB. Moist forest, especially along streams and waterfalls where it may be gregarious; 1100–1200 m
USES. None recorded
CONSERVATION NOTES. Widely distributed; least concern (LC)

SYN. *Aspidium protensum* Sw. in J. Bot. (Schrad.) 1800, 2: 36 (1801)
Ctenitis protensa (Sw.) Ching in Sunyatsenia 5: 250 (1940); Tard. in Mém. I.F.A.N. 28: 134, t. 24, fig. 5,6 (1953) & in Humbert, Fl. Madag. 5e fam. 1: 341 (1958); Alston, Ferns W.T.A.: 71, fig. 14, c, d (1959)

2. **Triplophyllum troupinii** (*Pic.Serm.*) *Holttum* in K.B. 41(2): 243 (1986). Type: Congo-Kinshasa, near Yangambi, *Pichi Sermolli* 5286 (FT-Pic.Serm., holo.)

Terrestrial; rhizome long-creeping and slender, not seen for East African material. Fronds spaced; stipe stramineous to reddish, 50–100 cm long, dark base with sparse covering of short glandular hairs and few narrowly lanceolate dark brown scales 7 × 0.5 mm; lamina thin-textured, mid green above, yellowish green below, deltoid, 47–60 × 60–74 cm, 3–4-pinnatifid; rachis with adaxial groove; pinnae in ± 13 pairs, apical pinnae gradually decrescent, lowest four pinnae remote; basal pinna pair to 42 cm long, bearing 6–7 pairs of stalked pinnules and several pairs of sessile or adnate ones, the basal basiscopic pinnule pair to 20 × 6 cm; second pinna-pair 26–35 cm, with long basiscopic pinnule; basiscopic pinnules of other pinnae becoming more equal to acroscopic ones; ultimate segments to 5 mm wide, obtuse, margin entire or slightly crenate; venation not reaching margin; pinna-costae and costae densely covered in short, septate glandular hairs 0.05–0.7 mm long, later increasingly glabrous, adaxially veins and veinlets covered in hairs to 1 mm long, a few hairs on lamina near margin. Sori 2–7 per segment, median on acroscopic veinlet, ± 1 mm in diameter; exindusiate (Uganda) or with indusium to 1 mm in diameter (Cameroon); sporangium with 14 indurate cells. Spores monolete, oval, exospore indistinct (Uganda material) maybe bluntly-echinate with equatorial membrane 0.05 mm long, 0.04 mm wide. Fig. 3: 6–10.

UGANDA. Masaka District: Towa Forest, June 1935, *A.S. Thomas* 1347! & 1348!
DISTR. **U** 4; from Liberia in the west to Gabon, Congo-Brazzaville and Congo-Kinshasa
HAB. Moist primary forest; ± 1200 m
USES. None recorded
CONSERVATION NOTES. Widespread; least concern (LC)

SYN. *Nephrodium variabile* Hook., Spec. Fil. 4:140 (1862) pro parte excl. typ.
Ctenitis troupinii Pic.Serm. in Webbia 39: 23, fig. 7, 8 (1985)

3. **Triplophyllum fraternum** (*Mett.*) *Holttum* in K.B. 41: 253 (1986). Type: Madagascar, *Boivin* s.n. (W, lecto., designated by Pichi Sermolli, 1985)

Terrestrial; rhizome long-creeping, ± 4(–5) mm in diameter, covered in dark brown scales, the smaller tightly appressed and ± 3 mm long, the longer narrowly lanceolate and curved and twisted, 7–8 mm long, tipped by septate, glandular hair. Fronds spaced, 1.5–2 cm apart; stipe ± 20–(40) cm long, 2.5(–4) mm diameter at base, decreasing to 1.5–2.5 mm diameter near apex, with dense long scales at base, rest of stipe with dense glandular septate hairs less than 0.5 mm long; lamina thin to subcoriaceous, dark green adaxially, green to pale olive below, narrowly deltoid, 27–33 × 20–28 cm, 3-pinnate; rachis adaxially sulcate, abaxially with hairs as on stipe; pinnae ± 7 and a gradually decreasing pinnatifid apex, the basal pair longest, to 16 cm, and with the first few basiscopic pinnules up to 8-pinnate and to 7 cm long, other pinnules pinnatifid (first acroscopic) but gradually decreasing in size and dissection to slightly crenate; next two pairs of pinnae slightly stalked, next four pairs adnate, grading into ± five adnate pinnules; margins crenate near base, entire near apex; pale sinus tissue ± 0.2 mm deep; pinna-costae adaxially sulcate and green to stramineous, densely short-hairy; veins and veinlets almost invisible, not reaching margin. Sori median on acroscopic veinlet, 1.2 mm in diameter; indusium honey-brown, 1.2 mm in diameter, of thick-walled square cells, fringed moderately with thick, 2–3-celled glands. Spores monolete, 0.4 mm long, 0.3 mm wide, membranous fringe 0.1 mm wide. Fig. 3: 11–14.

TANZANIA. Lushoto District: East Usambara Mts, Amani, Dodwe stream, June 1970, *Faden* 70/273!; Iringa District: Udzungwe Mts, Sanje Falls, Nov. 1995, *de Boer et al.* 724!
DISTR. **T** 3, 7; Guinea, Sierra Leone, Liberia, Ivory Coast, Bioko, Principe, Cameroon; Madagascar
HAB. Moist forest; 900–1000 m (data from one specimen only)
USES. None recorded for our area
CONSERVATION NOTES. Widespread; least concern (LC)

SYN. *Aspidium fraternum* Mett. in Kuhn, Fil. Afr.: 132 (1868)
Ctenitis fraterna (Mett.) Tardieu in Notul. Syst. 14: 242 (1952) & in Mém. I.F.A.N. 28: 135, t. 26/1–2 (1953) & in Humbert, Fl. Madag. Pter. 1: 340 (1954)

NOTE. Our material belongs to var. *fraternum*; var. *elongatum* (Hook.) Holttum is confined to Principe.

A specimen from Uganda, **U** 3, Busoga District: Kyabuma, June 1915, *Dummer* 2595! matches *T. fraternum* in dissection, but is glabrous on costa and costules, as in *T. jenseniae*. Conceivably a hybrid?

The description given is based on the East African material cited. Holttum's description, based on West African and Madagascan material, differed slightly in the rhizome scales (4–5 mm long).

This species has been confused with *T. jenseniae* (C.Chr.) Holttum; among the characteristics of *T. fraternum* are that pinnae of the basal pair are not very much longer than the pair above, also the absence of buds and the fact that major axes bear glandular hairs adaxially (fide Peter J. Edwards)

4. **Triplophyllum vogelii** (*Hook.*) *Holttum* in K.B. 41(2): 249 (1986). Type: Bioko [Fernando Po], *Vogel* 250 (K!, holo.)

Terrestrial; rhizome not seen, described as long-creeping, slender; rhizome scales ovate, to 3 × 1.7 mm, acuminate. Fronds spaced, to 80 cm long; stipe reddish to dark brown, 24–47 cm long, base 5 mm in diameter, decreasing to ± 2 mm below blade, adaxially grooved, near base with narrowly triangular scales to 5 × 1 mm, twisted, otherwise with rather dense septate hairs to 1 mm long, sometimes ctenitoid; lamina thin, dark green above, mid green below, deltoid-pentagonal, 30–35 × 25–39 cm, 3-pinnate to 4-pinnatifid; rachis with hairs as on stipe; pinnae 7–10 on each side of the rachis, near apex gradually grading into decreasing pinnatisect apex; basal pinna pair longest and most divided, each pinna to 24 cm long and curving upward, basal basiscopic pinnule ± 13.5 cm, 2-pinnatifid; second pinna pair with each pinna < 20 cm long, straight, 1-pinnate or with the basal pinnules slightly divided; ultimate segments slightly crenate to almost entire; each segment with a vein with 1–3 lateral veinlets, not reaching margin; pinna-costa and costa with dense septate hairs of ctenitoid appearance often catenate, glandular, those near branching forks 1.5–2 mm long, smaller on veinlets, insignificant or absent at sinus; adaxially a few large hairs restricted to veins and veinlets. Sori up to 9(–11) per lobe, on veinlets near tip, median, ± 1 mm in diameter; indusium ± 1 mm in diameter, of small square cells, notch to the centre, with ± 8 septate, catenate hairs 0.25–0.45 mm long on margin and ± 8 more such hairs distributed over surface. Spores 0.5 mm long, 0.3 mm wide, features not well seen. Fig. 3: 15–18.

UGANDA. Masaka District: Sese, Fumve, Feb. 1933, *A.S. Thomas* 903! & Sese town forest, June 1935, *A.S. Thomas* 1345!
DISTR. **U** 4; widely distributed in West Africa and Congo; Madagascar? (fide Holttum)
HAB. Moist primary forest; ± 1200 m
USES. None recorded for our area
CONSERVATION NOTES. Widespread; least concern (LC)

SYN. *Aspidium vogelii* Hook., Ic.Pl. 10: t. 921 (1854)
Aspidium lanigerum Mett. in Kuhn, Fil. Afr.: 135 (1868). Type: Congo, *C. Smith* s.n. (C, holo.)
Ctenitis vogelii (Hook.) Ching in Sunyatsenia 5: 250 (1940); Alston, Ferns W.T.A.: 71 (1959)
Ctenitis lanigera (Mett.) Tard. in Notul. Syst. 14: 343 (1952) & in Mém. I.F.A.N. 28: 141, t. 25/3–4 (1953); Alston, Ferns W.T.A.. 71, fig. 14/a–b (1959), as *C. lanuginosa*

NOTE. The characteristic appearance of this blade depends on the lack of curves: the straight (stiff?) pinna-costae are at an angle to the rachis; the first pinna pair rises parallel to those above and is the only one that curves rather sharply upward in the top-half of its length. The second distinctive feature is the compact, narrow triangle formed from pinnae above the second pinna pair.

3. LASTREOPSIS

R.C.Ching in Bull. Fan. Mem. Inst. Biol. 8, 4: 157 (1938); M.D. Tindale in Contr. New South Wales Nat. Herb. 3 (5): 249–339 (1965)

Polystichum § *Parapolystichum* (Keyserl.) C.Chr., Monogr. *Dryopteris* 2: 93 (1920)
Parapolystichum (Keyserl.) Ching in Sunyatsenia 5: 239 (1940)

Rhizome long- or short-creeping, rarely erect. Fronds spaced; stipe with adaxial groove, linear aerophores along ridges bordering groove; blade in almost all species decompound, deltoid-pentagonal, finely divided as in *Ctenitis*, but differing in scales which resemble those of *Tectaria* and in thickened basiscopic margins on leaflets which form wings on the axes to which they are attached; lowest pair of primary pinnae strongly basiscopically produced, catadromous below, anadromous towards tip of blade; main rachis bordered above by two prominent ridges, continuations of thickened leaf margins of pinnae; intervening broad shallow channel filled with ctenitoid hairs (short, articulated unbranched hairs); costae raised, veins free; ultimate leaflets with margins thickened and decurrent to join margins of groove of pinna-rachis, prominent costae of distal parts of pinnae decurrent to form raised band between thickened margins of groove. Sori orbicular, indusia present or fugacious; sporangia lacking appendages but usually bearing glands on stalk; spores winged or with many obtuse protuberances.

About 35 species, pantropical.

In both species as well as in *Megalastrum* an additional acroscopic pinnule is inserted between the first two ± opposite pinnule pairs of the basal pinna.

Lamina herbaceous, with bluish-grey metallic sheen above, to 60 × 60 cm; lamina glabrous adaxially but for dense brush-like arrangement of glandular trichomes ± 0.5 mm on costa and in grooves of larger axes; abaxially almost entirely glabrous . . . . 1. *L. currori*
Lamina coriaceous, metallic sheen absent, to 145 × 100 cm; veinlets and lamina abaxially with stout whitish hairs and tiny globular iridescent glands 0.2–0.3 mm . . . . 2. *L. vogelii*

1. **Lastreopsis currori** (*Mett.*) *Tindale* in Vict. Nat. 73: 184 (1957) & in Vict. Nat. 73: 282 (1965) & in Contrib. New S. Wales Nat. Herb. 3 (5): 282 (1965). Type: Cameroon, Efulen, 5 June 1895, *Curror* s.n. (K!, holo.; B, iso.)

Terrestrial or epiphytic; rhizome ascending to erect, more than 10 cm high and up to 3.5 cm wide; rhizome scales dark brown, narrowly lanceolate, 5–10 × 0.5–1 mm, with a few marginal processes. Fronds tufted, gemmiferous near apex of rachis and sometimes near apex of pinna rachis; stipe light brown to yellowish, 20–90 cm long, 1.5–6 mm in diameter at base, glabrous except for scales near rhizome and for the apex of the groove which is tomentose; lamina papyraceous, dark green above and often with bluish metallic sheen, paler beneath, quinquangular, 20–60 × 15–60 cm, 3–4-pinnate; rachis winged near apex, lower surface with many orange or yellow glandular hairs, a few ctenitoid hairs and some fibrillose scales, adaxial channels of vascular elements with dense mainly 3-celled ctenitoid hairs; pinnae in 14–18 pairs, the lowest pair much enlarged and basiscopically produced, 15–55 × 12–27 cm, with 12–20 pairs of pinnules; secondary rachises with many glandular hairs on lower surface and a few ctenitoid hairs, the last more common towards the apex; ultimate segments apex broadly rounded, margin crenulate; veins free, forked or pinnate in ultimate segments, not reaching the apex, catadromous throughout or almost so, often densely hairy. Sori median to submarginal on the acroscopic branch of minor veinlets, orbicular, 0.5–1.2 mm in diameter, indusiate; indusium fugacious, prominent to almost vestigial, reniform-orbicular, 0.3–0.7 mm in diameter, margin erose or fimbriate, sometimes with glandular hairs; sporangium stalk occasionally with oblong glandular hair (paraphysis). Spores monolete, pale yellow-transparent, exospore of inflated tubercles. Fig. 4: 7–15.

UGANDA. Kigezi District: Ishasha Gorge, Apr. 1998, *Hafashimana* 533!; Masaka District: Sese Islands, Fumua, June 1950, *G.H.S. Wood* 460!; Mengo District: Kyagwe, 1 km N of Maigwe, Sep. 1969, *Faden et al.* 69/1047!
TANZANIA. Morogoro District: N Uluguru Mts, Mkungwe Hill, July 1970, *Faden et al.* 70/377!
DISTR. **U** 2, 4; **T** 6; from Liberia to Cameroon, Congo-Kinshasa; Madagascar
HAB. Moist forest, sometimes near water or in swampy sites; may be locally common; 900–1500 m
USES. None recorded for our area
CONSERVATION NOTES. Widespread; least concern (LC)

SYN. *Aspidium currori* Mett. in Kuhn, Fil. Afr.: 130 (1868)
Dryopteris currori (Mett.) O.Ktze., Rev. Gen. Pl.: 812 (1891)
Ctenitis currori (Mett.) Tardieu in Notul. Syst. 14: 342 (1952)

NOTE. Our material belongs to subsp. *currori*; subsp. *eglandulosa* Tindale is confined to Congo-Kinshasa.

2. **Lastreopsis vogelii** (*Hook.*) *Tindale* in Contrib. New S. Wales Nat. Herb. 3 (4): 245 (1963) & in Contrib. New S. Wales Nat. Herb. 3 (5): 285 (1965). Lectotype: Bioko [Fernando Po], *Vogel* 229 (K!, lecto.), chosen by Alston, 1959)

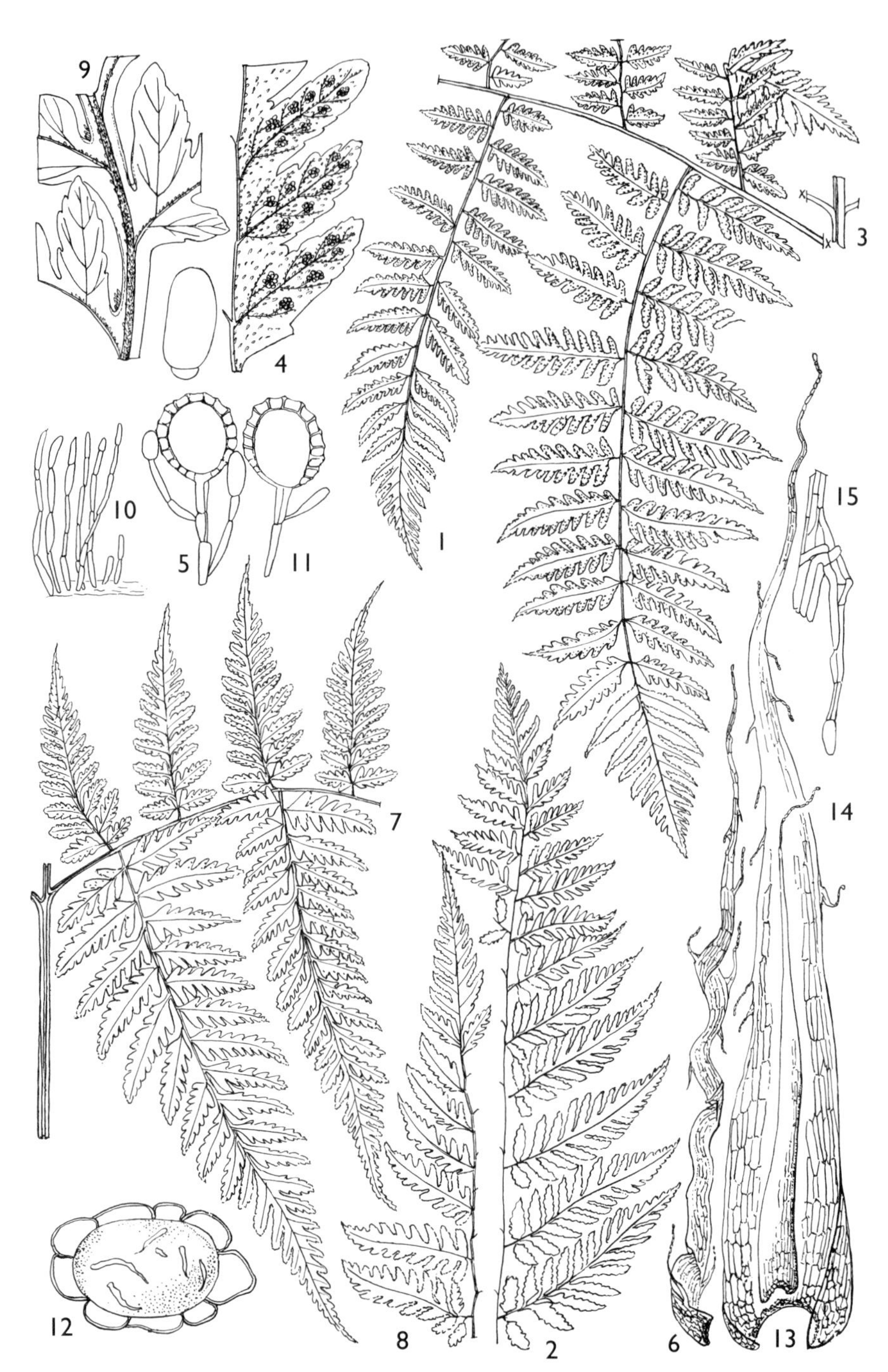

FIG. 4. *LASTREOPSIS VOGELII* — **1**, part of basal pinna, × 0.4 ; **2**, frond apex, × 0.4; **3**, base of rachis, × 0.4; **4**, part of pinnule, × 3.5; **5**, sporangium, × 100 ; **6**, stipe scale, × 8. *LASTREOPSIS CURRORI* — **7**, part of basal pinna, × 0.4; **8**, frond apex, × 0.4; **9**, pinnule part, adaxial, × 4; **10**, trichomes from costa, × 100; **11**, sporangium, × 100 ; **12**, spore, × 150; **13**, **14**, stipe scales, × 8; **15**, glandular trichome from scale margin, × 25. 1–6 from *Faden et al.* 69/968; 7–15 from *Poulsen et al.* 766. Drawn by Monika Shaffer-Fehre.

Terrestrial; rhizome ascending, with linear scales to 10 × 0.5 mm with long marginal processes. Fronds tufted, gemmiferous; stipe light brown to stramineous, to 65 cm long, glabrous except for a few scales near the base and a few dark brown scales in the grooves on the upper surface; lamina herbaceous to subcoriaceous, dark green, quinquangular, to 145 × 100 cm, 3-pinnate to 4-pinnatifid; rachis winged near apex, stramineous; pinnae (not counted) with the lowermost pair much enlarged and basiscopically produced, 30–52 × 25–30 cm; pinna rachis tomentose on upper surface, glabrous on lower surface; ultimate segments oblique, narrowly oblong, apex acute to obtuse; veins free, not reaching margin, catadromous except near apex of pinnules, with few stiff, spreading hairs on lower surface, costae with 3–4-celled brownish appressed hairs; lamina adaxially glabrous, abaxially both vascular system and lamina covered in stout white hairs and tiny glands. Sori median on the middle of simple veinlets or on acroscopic branch of forked veinlets, superficial, orbicular, 1–1.8 mm in diameter, exindusiate; stalked paraphyses from soral stalk, glandular head ovoid, sporangial stalk short with yellow, septate glandular hair (paraphysis). Spores monolete. Fig. 4: 1–6.

UGANDA. Masaka District: Sesse Islands, Buddu, 1903, *Dawe* 22!; Mengo District: Ssezzibwa Falls, 10 km W of Lugazi, Sep. 1969, *Faden & Evans* 69/968!
DISTR. **U** 4; Nigeria, Bioko, Cameroon, Congo-Kinshasa
HAB. Riverine and swamp forest; 1130–1200 m
USES. None recorded for our area
CONSERVATION NOTES. Widespread; least concern (LC)

SYN. *Polypodium vogelii* Hook., Sp. Fil. 4: 271 (1862)
Phegopteris vogelii (Hook.) Kuhn, Fil. Afr.: 124 (1868)
Dryopteris subcoriacea C. Chr., Ind. Fil.: 295 (1905). Type as for *Polypodium vogelii*
Ctenitis subcoriacea (C.Chr.) Alston in Bol Soc. Brot. 30, ser. 2a.: 12 (1956)

NOTE. Christensen based *Dryopteris subcoriacea* directly on *Polypodium vogelii* Hook., as a new name was necessary in *Dryopteris* due to the presence of *D. vogelii* (Hook.) C.Chr., based on *Aspidium vogelii* Hook. (1854). The lectotype for *D. subcoriacea* is therefore the same as for *Polypodium vogelii*.

4. CTENITIS

(C.Chr.) C.Chr. in Verdoorn, Man. Pteridol.: 543 (1938)

Dryopteris Adans. subgen *Ctenitis* C.Chr., Biol. Arb. Warming: 77 (1911)

Medium sized to large ferns, terrestrial or epilithic. Rhizome erect, even trunk-like, with clathrate scales. Fronds tufted; stipes with adaxial groove and basal scales; blade oblong-lanceolate, pinnate-pinnatifid to tri-pinnate, catadromous, maybe anadromous at very base; penultimate divisions often oblique with unequal base, decurrent and wing-connected, margin entire, serrate, crenate to pinnatifid; veins free, simple or forked, basal vein of a segment rarely springing from costa instead of costule; indumentum of narrow paleae that proximally bear short marginal outgrowths and often unicellular glands, pluricellular and acicular hairs on axes and adaxial lamina, surfaces occasionally with ctenitoid hairs. Sori small circular, median, on enclosed veinlets, indusiate; sporangium short-stalked.

About 150 species in the tropics and south temperate area; absent from Australia

Ctenitis cirrhosa (*Schumach.*) *Ching* in Sunyatsenia 5(4): 250 (1940); Alston, Ferns W.T.A.: 70 (1959); Schelpe, F.Z. Pterid.: 232, t. 543 (1970); Jacobsen, Ferns Southern Afr.: 453 (1983); Faden in U.K.W.F. ed. 2: 35 (1994). Type: 'Guinee', *Thonning* 302 (C, holo., 3 sheets)

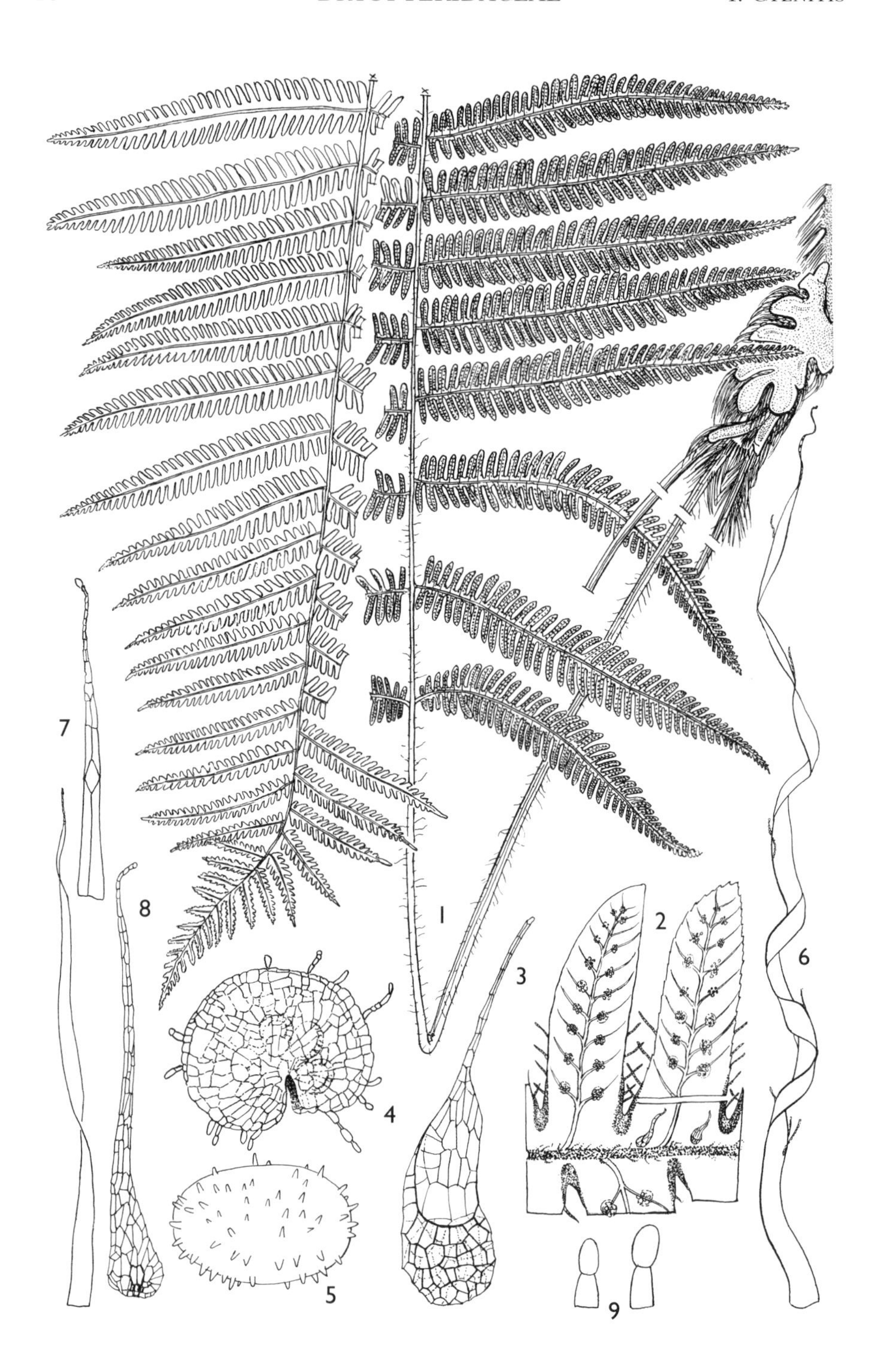

FIG. 5. *CTENITIS CIRRHOSA* — **1**, habit, × 0.4 ; **2**, pinnule detail, × 3; **3**, bullate scale from rachis, × 12; **4**, indusium, × 50; **5**, spore, × 200 ; **6**, twisted scale, × 90; **7**, glandular scales, × 15; **8**, scale, × 40; **9**, glandular hairs, × 90. 1–8 from *Faden et al.* 77/887; rhizome from *Poulsen* 1017 and *Cribb* 11152. Drawn by Monika Shaffer-Fehre.

Terrestrial; rhizome erect or ascending, rarely short-creeping; rhizome scales narrowly lanceolate, 7–20 × 1–1.2 mm, with few marginal gland-tipped protuberances. Fronds tufted, erect; stipe dark at base, stramineous above, 20–45 cm long, 5 mm across at base, initially densely covered in soft brittle linear-lanceolate scales 3–10 × 0.1 mm, soon lost from upper stipe, scales near base like those of rhizome; lamina deltoid or more often ovate-elongate, 25–100 × 30–40 cm, 2-pinnate; rachis not winged, indument like stipe; pinnae in 15–20+ pairs, the lowermost slightly smaller than the next pair up, minutely stalked (1–2 mm), the longest to 20 × 4 cm, lamina apex gradually decrescent with more and more adnate segments; pinnules up to ± 40 per pinna, very slightly falcate, sterile pinnules almost straight-sided, 20 × 4–5 mm, tip obtuse, fertile pinnules spathulate, to 17 × 3 mm; all margins with sparse transparent ± 0.8 mm long catenate hairs; pinna costa and costule densely beset with glandular hairs 0.3–0.9 mm long, tipped with orange gland that can be lost (i.e. diagnostic ctenitoid hairs); costule and to lesser extent veinlets and margin of pinnule covered in straight, 0.2 mm long, unicellular, blunt white hairs. Sori median, more numerous or persistent on acroscopic side of pinnule, ± 0.8 mm diameter; indusium oval, 0.5 × 0.7 mm diameter, notch to centre, few 2-celled glands on surface and margin and few 4–5-celled trichomes extending from margin, indusium when drying withdrawn into centre of sorus. Spore pale yellow, 0.1 × 0.1mm, echinate, spines 0.01 mm, blunt, few. Fig. 5.

Uganda. Toro District: Kibale National Park, near Kanyawara, July 1994, *Poulsen et al.* 669!; Busoga District: White Nile at Kibibi, Feb. 1953, *Wood* 624!; Mengo District: Mabira Forest, Bwola, July 1971, *Katende* 1163!

Kenya. N Kavirondo District: Kakamega Forest, NE of Forest Station, Nov. 1969, *Faden et al.* 69/1981! & idem, along Kubiranga stream, Mar. 1977, *Faden & Faden* 77/887!; Teita District: Taita Hills, Mwabirwa Forest Station, May 1985, *Kabuye et al.* 807!

Tanzania. Lushoto District: Amani, Dodwe stream, June 1970, *Faden* 70/274!; Ulanga District: Ngongo Mt, Jan. 1979, *Cribb et al.* 11152!; Lindi District: Rondo Forest Reserve, Feb. 1991, *Bidgood et al.* 1562!

Distr. **U** 2–4; **K** 5, 7; **T** 3, 6, 8; throughout the tropics and south temperate zones

Hab. Moist forest, often near streams; may be locally common; (650–)900–1700 m

Uses. None recorded for our area

Conservation notes. Widespread; least concern (LC)

Syn. *Aspidium cirrhosa* Schumach. in Vid. Selsk. Afh. 4: 231, 232 (1829)
Nephrodium cirrhosum (Schumach.) Baker in Hook. & Bak., Syn. Fil.: 472 (1868)
Dryopteris cirrhosa (Schumach.) Kunze, Rev. Gen. Pl. 2: 812 (1891)

Note. Possible dimorphism exists among blades of the same gathering where the sterile blade appears larger, more compact, by virtue of touching or overlapping pinnae and by wider pinnules. For instance: sterile blade, 40 × 32 cm with ± 17 pinna pairs below tip and fertile blade 62 × 32 cm wide with ± 24 pinna pairs. Plants from Kenya appear to have larger blades than those from Uganda or Tanzania.

Interval ('Ctenitis' derived from the Greek ktena = comb) between fertile pinnae is 2 cm at base of blade to 5 mm and 2 mm near tip.

5. **HYPODEMATIUM**

Kunze in Flora 16: 690 (1833)

Rhizome creeping with dense lanceolate-acuminate rhizome-scales. Fronds tufted; stipe not articulated, with a tuft of scales at the base. Lamina ovate-triangular to pentagonal in outline, herbaceous, with the lowest pinnae basiscopically much developed, 3–4-pinnatifid, pilose with needle-like unicellular hairs; veins free. Sori subcircular; indusia reniform, pilose.

An Old World genus of about 3 species, with only one in Africa.

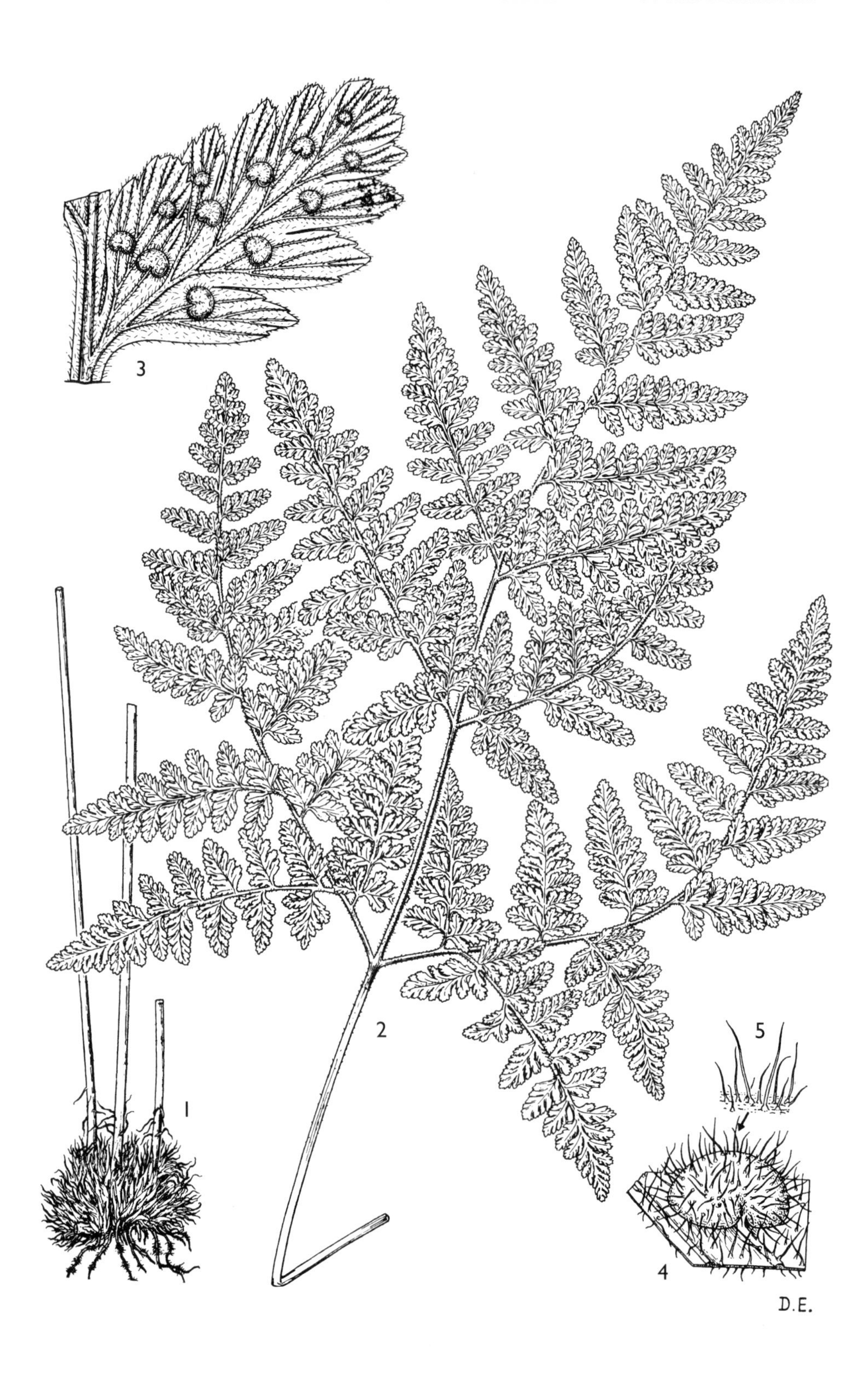
3
2
1
5
4
D.E.

Hypodematium crenatum (*Forssk.*) *Kuhn* in Von der Decken, Reisen, Bot. 3, 3: 37 (1879); Tardieu, Fl. Madag. 5 (1): 327 (1958); Schelpe, F.Z., Pterid.: 230, t. 66 (1970) & in Expl. Hydrobiol. Bassin L. Bangweolo & Luapula 8 (3) Ptérid.: 90, fig. 28 (1973); Faden in U.K.W.F.: 48 (1974); Kornaś, Distr. Ecol. Pterid. Zambia: 105, fig. 69a (1979); Jacobsen, Ferns S. Afr.: 451, t. 341, map 168 (1983); Schelpe & N.C. Anthony, F.S.A., Pterid.: 263, map 229 (1986); Burrows, S. Afr. Ferns: 320, t. 53/5, fig. 324, 324a, map (1990); Faden in U.K.W.F. ed. 2: 33 (1994). Type: Yemen, Bolghose, *Forsskål* s.n. (type lost?)

Terrestrial; rhizome short, with tufted fronds and with dense ferrugineous concolorous entire lanceolate-acuminate scales ± 1 cm long. Fronds tufted, arching; stipe up to 12 cm long, stramineous, glabrous except for scales, similar to those on the rhizome, mostly in a dense basal tuft; lamina softly herbaceous, up to 33 × 30 cm, ovate-triangular in outline, 4-pinnatifid with the basal pinnae developed basiscopically; rachis stramineous, pilose; pinnae up to 22 cm long, oblong-acute in outline towards the apex, unequally triangular-ovate at the base; pinnules of the upper pinnae and basiscopic pinnule-segments of the lowest pinnae up to 2 × 0.9 cm, deeply pinnatifid into oblong obtuse crenate lobes, pilose with unicellular hairs on both surfaces. Sori subcircular, up to 12 per pinnule (or pinnule-segment of basal pinnae), 1–1.5 mm in diameter; indusium reniform, pilose with straight white unicellular hairs. Fig. 6.

KENYA. South Kavirondo District: Kanam, Homa Mt., Dec. 1934, *Allen Turner* in CM 3632!

DISTR. **K** 5; Congo-Kinshasa, Sudan, Ethiopia, Angola, Zambia, South Africa; Madagascar, Mascarene Is.; Arabia; Cape Verde Is., India, Malaya, Philippines, Indochina, China, Japan and Polynesia

HAB. Probably rocky places in bushland; ± 1200 m

USES. None recorded for our area

CONSERVATION NOTES. Widespread; least concern (LC)

SYN. *Polypodium crenatum* Forssk., Fl. Aegypt.-Arab.: CXXV, 185 (1775)

Aspidium odoratum Willd., Sp. Pl. ed. 4, 5: 286 (1810). Type: Mauritius, *Bory de St. Vincent* s.n. (B-W 19833, holo.; microfiche!)

Nephrodium hirsutum D.Don, Prodr. Fl. Nepal.: 6 (1825). Type: Nepal, "alpibus", *Wallich* s.n. (BM!, holo.)

Cystopteris odorata (Willd.) Desv. in Mém. Soc. Linn. Paris 6 (2): 264 (1827)

Hypodematium rueppelianum Kunze, Farnkr. 1: 41, t. 21 (1840). Type: Ethiopia, "Abyssinian mountains", *Rüppell* s.n. (FR, syn.) & *Schimper* 358 (? LZ, syn.)

Lastrea hirsuta (D.Don) Moore, Ind. Fil.: 88 (1857)

Aspidium eriocarpum Mett. in Abh. Senckenb. Naturf. Ges. 2: 344 (1858) & in Mett., Farngatt. 4: 60 (1859) ("1858"). Types: Nepal, *Wallich* 342 (K!, K-Wall!, syn.)[1]

Lastrea odorata (Willd.) Wawra in Oest. Bot. Zeitschr. 13: 144 (1863)

Nephrodium odoratum (Willd.) Baker, Syn. Fil.: 280 (1868)

Aspidium crenatum (Forssk.) Kuhn, Fil. Afr.: 129 (1868)

Lastrea crenata (Forssk.) Bedd., Ferns Brit. Ind., Suppl.: 18 (1876)

Nephrodium crenatum (Forssk.) Baker, Fl. Maurit.: 49 (1877)

Dryopteris crenata (Forssk.) Kuntze, Rev. Gen. Pl. 2: 811 (1891); Sim, Ferns S. Afr. ed. 2: 111, t. 22 (1915)

NOTE. An incredibly erratic species known from single widely scattered localities throughout its range in tropical Africa, apparently only found once in the Flora area.

[1] Mettenius cited Nepal, *Wallich* 342 (sphalm. 324), Himalaya, *Hofmeister* and Ethiopia, *Schimper*. Schelpe in F.Z. gives 'type' from Nepal! There are 3 *Wallich* syntypes in K and 2 in K-WALL.

FIG. 6. *HYPODEMATIUM CRENATUM* — **1**, rhizome and stipe bases, × $^1/_2$; **2**, frond, × $^1/_2$; **3**, pinnule, lower surface, × 10; **4**, sorus, × 18; **5**, indusium hairs, × 18. 1–5 from *Schimper* 617. Drawn by Derek Erasmus.

6. **MEGALASTRUM**

Holttum in Gard. Bull. Straits Settlements 39: 161 (1987)

Plants terrestrial; rhizome erect, massive. Fronds tufted, large, fertile more open than sterile; stipes and rachis with adaxial groove, discontinuous with grooves of lower order axes; blade pinnately compound, pinnae deeply lobed; venation free, simple or forked ending close to or in margin, both anadromous and catadromous; indumentum of ctenitoid hairs present in most species, in some species with pluricellular, curved, rigid, ± acicular hairs, and small short and long unicellular glandular hairs, uniseriate cylindrical hairs and acicular hairs along axes and adaxial lamina. Sori small, circular, on vein branches; sporangium long-stalked, 13–15 indurated cells, indusium reniform with unicellular glands, cylindrical & acicular hairs superficially and along margins. Spores monolete, fusiform with blunt tips, echinate, 40–50 mm long.

Approximately 30 species in the Neotropics; one of these extending to Africa and the Madagascan region.

Megalastrum lanuginosum (*Kaulf.*) *Holttum* in Gard. Bull. Strait Settlements 39: 161 (1987); Faden in U.K.W.F. ed. 2: 35 (1994). Type: Mauritius, *Petit-Thouars* s.n. (B-W 19808, holo.)

Terrestrial; rhizome erect or ascending, ± 15 cm high, to 5 cm in diameter; rhizome scales golden brown, linear, to 25–30 mm long, margin serrulate. Fronds tufted, 1.3–1.5 m long; stipe dark auburn near base, stramineous or green above, 64–100 cm long, at base ± 1 cm in diameter, 0.5 cm below blade, near base with dense ferruginous scales as on rhizome; lamina membranaceous, dull green above, pale green below, ovate to deltoid, 66–130 × ± 80 cm, tripinnate-pinnatisect, basally 4-pinnatifid; number of pinnae not clear, but more than 15, basal pinnae the longest and up to 40 × 14–27 cm, stalked, ± 18 pinnule pairs, strongly basiscopically developed, first basiscopic pinnule to 20 × 7(–9) cm with ± 19 pairs of 2nd order pinnules; median pinnae ± 30 × 9 cm, with up to 32 pinnules, the lower pinnate, the upper lobed; ultimate segments oblong, rounded, ± 3 mm wide, crenate, obtuse; lamina lightly pubescent on both surfaces, more densely on costa and veins, indument of septate hairs and few scales. Sori median, most often on acroscopic veinlet, 0.4–1 mm in diameter; indusium circular, ± 1 mm in diameter, deep notch at point of attachment, tissue of thick red-brown radial walls, with paler sometimes flabellate margins, persistent hairs and marginal glands, finally withdrawn into centre of sorus; sporangia dark brown. Spores monolete, 0.1 mm long, 0.08 mm wide, uniformly spinulose. Fig. 7.

KENYA. Kiambu District: Kikuyu Escarpment Forest, junction of Kitikuyu and Karamenu Rs., Apr. 1971, *Faden et al.* 71/282!; Meru District: Ithagune Forest, 14 km from Nkubu, June 1969, *Faden et al.* 69/763!; Teita District: Taita Hills, Ngangao, Dec. 1998, *Luke et al.* 5499!

TANZANIA. Lushoto District: Shume-Magamba Forest Reserve, May 1987, *Kisena* 459!; Morogoro District: Uluguru Mts, Mwere valley, Sep. 1970, *Faden et al.* 70/585!; Iringa District: Mwanihana Forest Reserve above Sanje, Oct. 1984, *D.W. Thomas* 3862!

DISTR. **K** 4, 7; **T** 3, 6, 7; Bioko, São Tomé, Malawi, Mozambique, Zimbabwe and South Africa; Madagascar, Mauritius

HAB. Moist forest, often near water; may be locally common; (1000–)1400–2400 m

USES. None recorded for our area

CONSERVATION NOTES. Widespread; least concern (LC)

SYN. *Aspidium lanuginosum* Kaulf., Enum.: 244 (1824)
Nephrodium lanuginosum (Kaulf.) Desv., Prodr.: 262 (1827)
Lastrea lanuginosa (Kaulf.) T.Moore, Ind.: 87 (1858)
Polystichum lanuginosum (Kaulf.) Keyserl., Pol. Cyath. Hb. Bung.: 45 (1873)

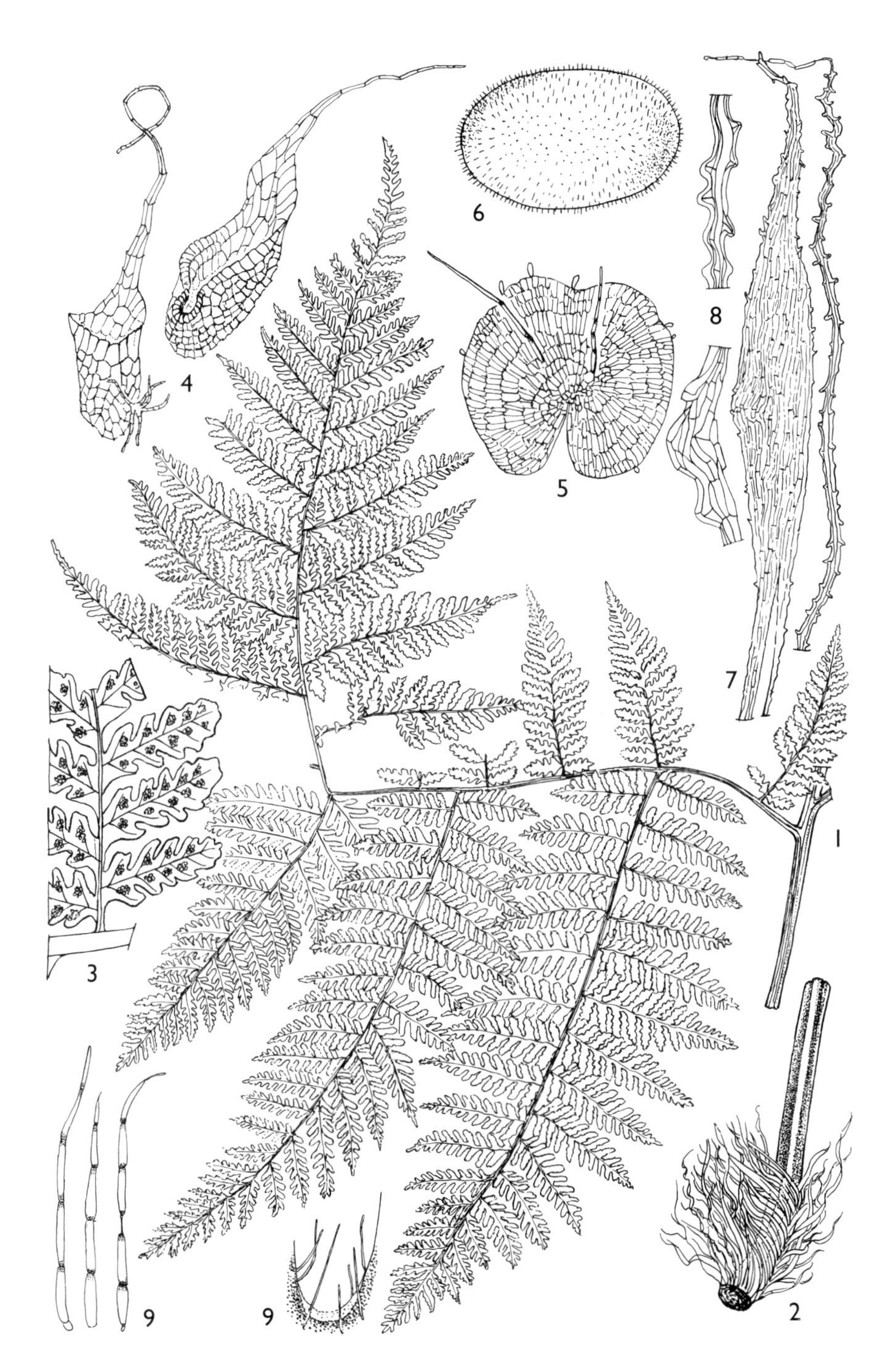

FIG. 7. *MEGALASTRUM LANUGINOSUM* — **1**, lowermost pinna, × 0.4 ; **2**, part of rhizome, × $^1/_2$; **3**, pinnule detail, × 1.5; **4**, costa scales, × 40; **5**, indusium, × 30 ; **6**, spore, × 100; **7**, rhizome scale, × 3; **8**, scale margin detail, × 100; **9**, stipe scales, × 3. 1–3, 6–8 from *Faden et al.* 71/282; 4 from *Faden et al.* 69/504; 5 from *Faden et al.* T468; 9–10 from *Luke et al.* 5499. Drawn by Monika Shaffer-Fehre.

Dryopteris lanuginosa (Kaulf.) C.Chr., Ind.: 273 (1906); Sim: 110 (1915)
Ctenitis lanuginosa (Kaulf.) Copel., Gen. Fil.: 124 (1947); Schelpe, F.Z., Pter.: 232, t. 67 (1970); Burrows, S. Afr. Ferns: 324, t. 54.1, fig. 78/327, map (1990)

NOTE. Observed in most specimens an additional acroscopic pinnule between first two basiscopic ones (cf. both species of *Lastreopsis*).
Lamina has odour of cut grass when crushed (*fide* Faden).

7. DIDYMOCHLAENA

Desv. in Mag. Ges. Naturf. Fr. Berlin 5: 303 (1811)

Rhizome erect, forming a short caudex, with large sub-entire brown rhizome-scales. Fronds tufted; stipe not articulated, scaly. Lamina oblong-ovate, 2-pinnate, firmly herbaceous; pinnules dimidiate-trapeziform, subsessile, articulate; veins free. Sori broadly elliptic, borne towards the acroscopic margin of the pinnules; indusium peltate with a narrow elongate stalk attached by a median linear ridge.

Pantropical genus with few species; usually considered monotypic but some Madagascan taxa seem distinct.

Didymochlaena truncatula (*Sw.*) *J.Sm.* in Hook., Lond. Journ. Bot. 4: 196 (1841); V.E. 2: 14, fig. 9 (1908); Tardieu, Fl. Madag. 5 (1): 304 (1958); Alston, Ferns W.T.A.: 69 (1959); Tard., Fl. Cameroun 3: 254, t. 38/68 (1964); Schelpe, F.Z., Pterid.: 220, t. 64/E (1970); Faden in U.K.W.F.: 45 (1974); Schelpe & Diniz, Fl. Moçamb., Pterid.: 235 (1979); Jacobsen, Ferns S. Afr.: 432, t. 325, map 158 (1983); Pic. Serm. in B.J.B.B. 55: 176 (1985); Schelpe & N.C. Anthony, F.S.A., Pterid.: 243, fig. 83/1, map 211 (1986); Burrows, S. Afr. Ferns: 299, t. 50/3, fig. [304], map (1990); Faden in U.K.W.F. ed. 2: 35 (1994). Type: Java?, Houtt., Nat. Hist. 14: t. 100, fig. 1 (1783) (lecto.)

Terrestrial; rhizome up to 2.5 cm in diameter, erect, forming a short woody caudex up to 20 cm high and 15 cm in diameter with attenuate subentire pale to dark brown rhizome-scales up to 2 × 0.1 cm with a few filamentous marginal outgrowths. Fronds tufted, 0.7–2.1 m tall but up to 3.6 m long, arching, firmly herbaceous; stipe stramineous, up to 50 cm tall, grooved, set with a mixture of narrowly ovate to linear-lanceolate brown scales up to 3 × 0.4 cm; lamina up to 147 × 48 cm, oblong-ovate, 2-pinnate, not reduced at the base; pinnae in about 20–40 pairs up to 25 × 4 cm, very narrowly oblong, acuminate, pinnatifid into up to 35(–40) pairs of dimidiate oblong-rhombic petiolate pinnules up to 2.5 × 1 cm, the basiscopic margin entire, thickened, with between 1–5 septate, ± stiff bristles (1(–1.7)–2 cm long), the acroscopic margin crenate (fertile pinnules) to serrate (sterile pinnules), glabrous on both surfaces at maturity; rachis and secondary rachises stramineous with persistent pale-brown scales similar (but shorter) to those on the stipe. Sori 1–11 per pinnule, borne in a slight depression nearer the acroscopic margin than the midrib, ± massive, broadly elliptic, up to 2.5 × 2 mm; indusium dark-brown with a paler border, with a very narrow elongate stalk, broadly elliptic, up to 3.5 × 2 mm, entire. Fig. 8.

UGANDA. Ankole District: Kasyoha-Kitomi Forest, NE of Kyambura R., 7 June 1994, *Poulsen et al.* 536!; Kigezi District: Ishasha Gorge, 5 Aug. 1971, *Katende* 1269!; Elgon, near R. Nametaba, 28 Oct. 1916, *Snowden* 491!
KENYA. North Kavirondo District: 8 km E of Kakamega Forest Station, 18 Sept. 1949, *Maas Geesteranus* 6292!; Kericho District: 5 km E of Kericho, along R. Timbilil, 11 June 1972, *Faden et al.* 72/310!; Teita District: Kasigau Mt, 1 June 1969, *Gillett* 18764!
TANZANIA. Mbulu District: Marang Forest, 16 June 1967, *Vesey-Fitzgerald* 5282!; Mpanda District: Mahali Mts, Sisaga, 29 Aug. 1958, *Newbould & Jefford* 1934!; Iringa District: Mufindi, Kigogo R., 23 Mar. 1962, *Polhill & Paulo* 1841!

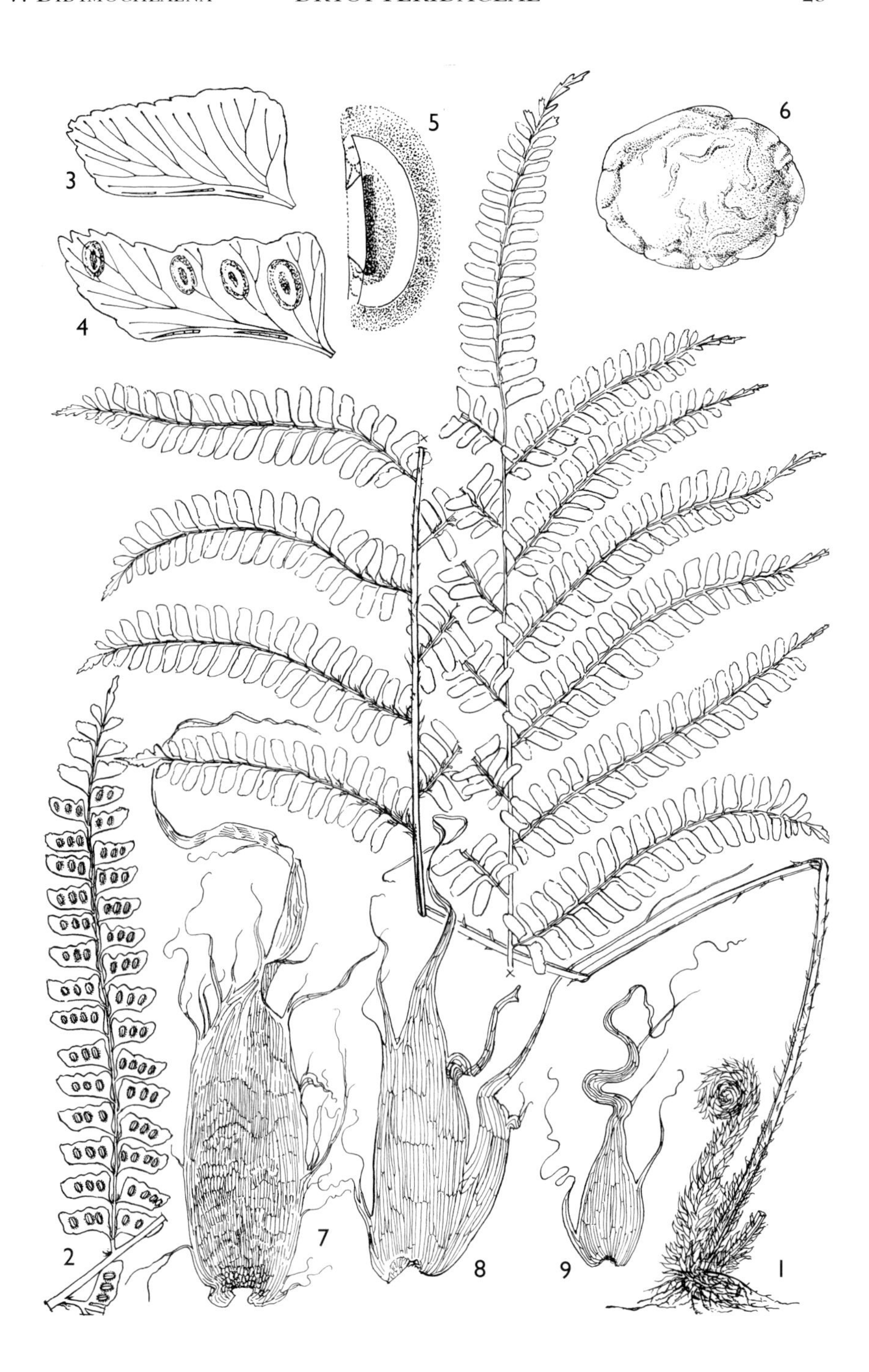

FIG. 8. *DIDYMOCHLAENA TRUNCATULA* — **1**, habit, × 0.4 ; **2**, fertile pinna, × 0.5 ; **3**, sterile pinnule, × 2; **4**, fertile pinnule, × 2; **5**, sorus section, × 6 ; **6**, spore, × 600; **7–9**, scales from rhizome and pinnule, × 4. 1–3, 6–9 from *Mtui & Sigara* 63; 4 from *Last* 3/1885. Drawn by Monika Shaffer-Fehre.

DISTR. **U** 2–4; **K** 4–7; **T** 2–4, 6–8; Bioko, São Tomé and Cameroon to Congo-Kinshasa, Burundi, Rwanda, Ethiopia, Angola, Malawi, Mozambique, Zimbabwe, South Africa; pantropical
HAB. Evergreen forest including mist and rain forest; 1000–2250 m
USES. None recorded for our area
CONSERVATION NOTES. Widespread; least concern (LC)

SYN. *Adiantum lunulatum* Houtt., Nat. Hist. 14: 209, t. 100, fig. 1 (1783), *non* Burm. f. (1768). Type as for *D. truncatula, nom. illeg.*
Aspidium truncatulum Sw. in Journ. Bot. (Schrad.) 1800, 2: 36 (1801)
Didymochlaena lunulata (Houtt.) Desv. in Mém. Soc. Linn. Paris 6: 282 (1827); Hieron. in E.J. 28: 341 (1900); F.D.O.-A.: 59 (1930), *nom. illeg.*
D. dimidiata Kunze in Linnaea 18: 122 (1844). Type: South Africa, *Gueinzius* s.n. (LZ†, holo., K!, iso., L, iso.)
Nephrolepis lunulata (Desv.) Keys., Pol. Cyath. Herb. Bung.: 40 (1873)
[*Didymochlaena microphylla* sensu Tard. in Mém. I.F.A.N. 28: 153 (1953), *non* (Bonap.) Christensen]

NOTE. Tardieu-Blot has a var. *attenuata* from E Madagascar.

8. **CYRTOMIUM**

C. Presl, Tent. Pter.: 86, t. 2, fig. 26 (1836)

Kramer et al. in Kubitzki, Fam. Gen. Vasc. Pl. I: 114 (1990)

Terrestrial or epilithic. Rhizome short, suberect to erect. Fronds tufted; stipe and rachis with adaxial groove not opening to sulci of lower order axes. Lamina herbaceous to coriaceous. Blade 1-pinnate, narrowly oval, apex imparipinnate or pinnatifid, acuminate ± falcate and auricled on acroscopic side; veins anastomosing with simple excurrent included veinlets, the latter bearing sori; indumentum of narrow paleae occurring on rhizome and axes, also with simple centrally pluriseriate hairs occurring abaxially on lamina. Sori round, indusium large, peltate, persistent or caducous.

A temperate or tropical genus of 9 species, centred in eastern Asia. Until recently treated as *Phanerophlebia* Presl (Tryon & Tryon, 1982), a name now reserved for the neotropical species of the genus (Kramer et al. in Kubitzki, Fam. Gen. Vasc. Pl. I: 114 (1990)).

Phanerophlebia (America) and *Cyrtomium* (Asia, Hawaii) are closely related genera of the tropics, separated on account of their venation. Both genera were described by Presl; Tryon & Tryon (1982) treated the former as *Phanerophlebia*, but present literature largely retains *Cyrtomium* for the Old World genera. Kramer in Kubitzki does still only recognise *Polystichum*; he argues that *Cyrtomium* and *Phanerophlebia* have characters that are not sufficiently correlated or constant.

Cyrtomium caryotideum (*Hook. & Grev.*) *C.Presl*, Tent. Pterid.: 86, t. 2 fig. 26 (1836); Jacobsen, Ferns S. Afr.: 255 (1983); Schelpe & Anthony, F.S.A.: 251 (1986); J.P. Roux, Conspect. southern Afr. Pterid.: 131 (2001). Type not designated

Terrestrial, epiphytic or lithophytic; rhizome erect, to 6 cm high; rhizome scales shiny dark brown, ovate, 8–13 × 2–3 mm, acuminate, pale centre, pale fimbriate margins. Fronds tufted; stipe with narrow adaxial and lateral grooves, 8–29 cm long, with pale brown scales 8–15 × 1.5–3 mm, irregularly fimbriate, denser on sterile fronds, less dense on fertile fronds; lamina dark green adaxially, pale greyish green below, coriaceous or thin, narrowly ovate or elliptic, 20–40 × 8–18 cm, 1-pinnate; rachis abaxially with sparse fimbriate scales, denser around stalks of pinnae; pinnae 3–12 on each side of the rachis, irregularly rhomboid-falcate, 4.7–8.5 × 2–2.5 cm, base cuneate on basiscopic side with acroscopic side truncate and auriculate, apex

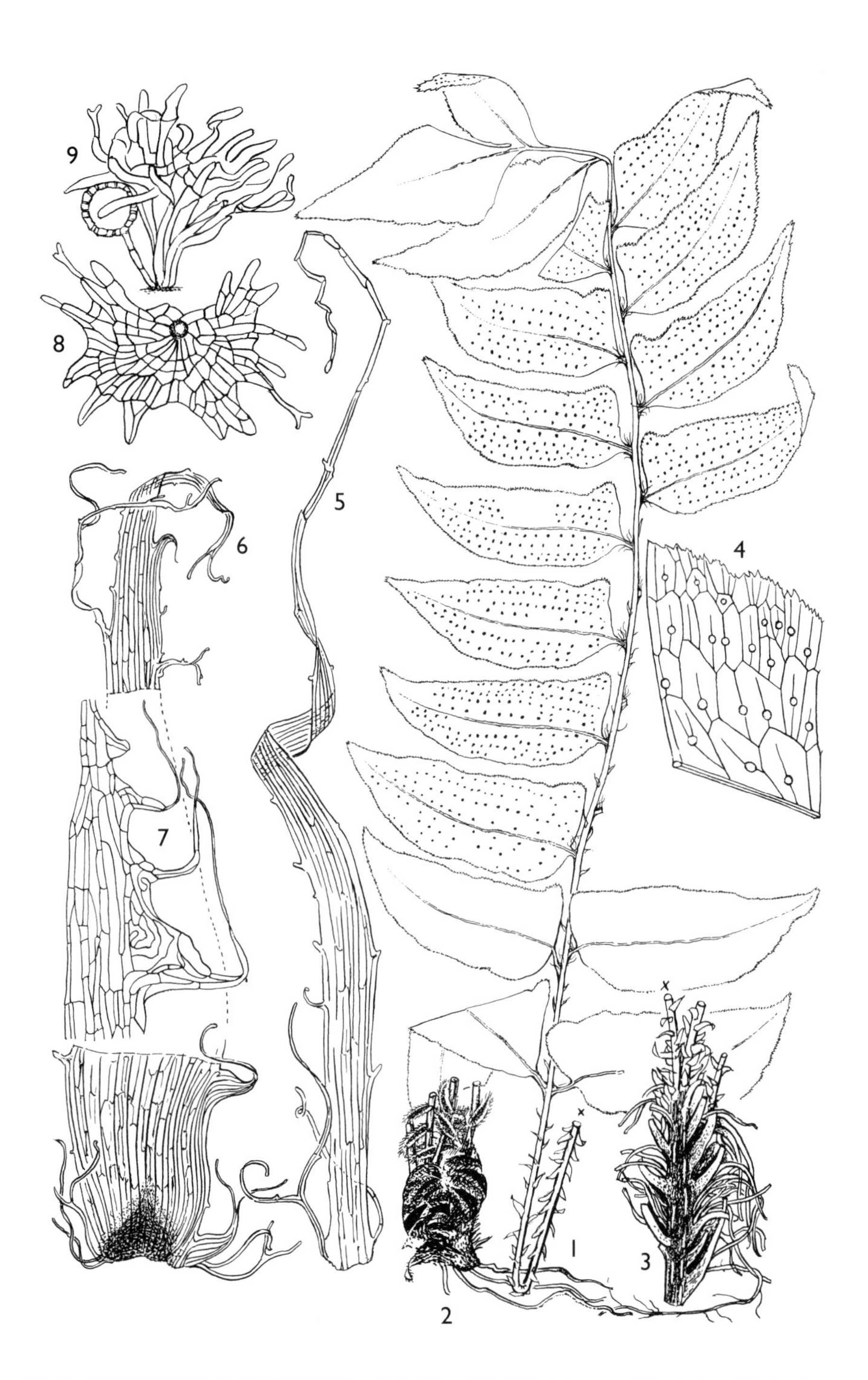

FIG. 9. *CYRTOMIUM CARYOTIDEUM* — **1**, habit, × 0.5; **2–3**, rhizomes (3 in l.s.), × 0.5; **4**, pinna detail, × 2; **5**, scale from stipe, × 20 ; **6**, rhizome scale upper part, × 20; **7**, scale margin, × 60; **8**, indusium, × 35; **9**, indusium lateral view, × 35. 1–2 from *Bidgood et al.* 541; 3, 5–7 from *Mtui & Sigara* 57; 4 from *Peter* 16148; 8 from *Pocs* 6444/T. Drawn by Monika Shaffer-Fehre.

up-turned, acuminate, blunt, margin finely denticulate or serrate particularly on acroscopic lobe; apical pinna triangular with unilateral basal lobe; veins anastomosing regularly, forming 2–3 ± parallel zones of areoles, traversed by pinnate veins, on periphery free veinlets reach serrate points but not margin. Sori ± regularly positioned on acroscopic free branch in areole, soral lines approximately parallel to longitudinal axis of pinna, 0.3–2 mm in diameter; indusium an ephemeral dark rust-red funnel-forming peltate stalk 0.25 mm, opening into highly dissected umbrella shape ± 1.8 mm in diameter; on dehiscing, peltate funnel-stalk detaches like plug from receptacle. Spores monolete, oval, 0.03 × 0.025 mm. Fig. 9.

var. **micropterum** (*Kunze*) *C.Chr.* in Am. Fern J. 20: 52 (1930) (as *micropteris*). Type: C.Presl, Tentamen pteridographiae t. 2, fig. 26 (1936), iconotype

Pinnae 6–12, with a single blunt auricle; terminal pinna equivalent to the other pinnae; lamina coriaceous with a serrate margin.

UGANDA. Ankole District: Bunyamguru, July 1939, *Purseglove* 859!
KENYA. Kericho District: Sotik, Kibajet Estate, Sep. 1949, *Bally* 7441! & Sotik, above Sisis R., May 1960, *Bally* 12255!; N Kavirondo District: Isiukhu R. on Kambiri–Vihiga road, Dec. 1969, *Faden & Rathbun* 69/2105!
TANZANIA. Kilimanjaro, hill behind Kilimanjaro Timbers, July 1993, *Grimshaw* 93/419!; Lushoto District: Camphor Forest, Sep. 1981, *Mtui & Sigara* 57!; Morogoro District: Nguru Mts above Maskati, Mar. 1988, *Bidgood et al.* 541!
DISTR. **U** 2; **K** 5; **T** 2, 3, 6; Ethiopia, Malawi; Madagascar to southern China and southern India
HAB. Moist forest, dry forest along river, plantation; may be terrestrial, or less often epiphytic or on dry shady rock; 1400–2100 m
USES. None recorded for our area
CONSERVATION NOTES. Widespread; least concern (LC)

SYN. *Aspidium caryotideum* Hook. & Grev., Icon. Fil. 1: t. 69 (1828)
Cyrtomium micropterum (Kunze) Ching, Icon. Fil. Sin. 3: t. 127 (1935) as '*micropteris*'; J.P. Roux, Conspect. southern Afr. Pterid.: 131 (2001)
Aspidium anomophyllum Zenker forma *micropteris* Kunze in Linnea 24: 278 (1851), as '*microptera*'
Phanerophlebia caryotidea (Hook. & Grev.) Copel. var. *micropteris* (Kunze) Tardieu, Fl. Madag. 1: 326 (1958); Jacobsen, Ferns S. Afr.: 455 (1983); Faden in U.K.W.F. ed. 2: 35 (1994)

NOTE. The typical variety differs in having 3–6 pinnae pairs, with 1–2 acute auricles; the terminal pinna is the largest; lamina thin, margin dentate with aristate teeth.

9. POLYSTICHUM

Roth, Tent. Fl. Germ. 3: 31 (1799); Roux, *Polystichum* in Africa: 1–375 (1998)

Terrestrial or lithophytic ferns, rarely epiphytic. Rhizome erect or creeping, with dense scales. Fronds tufted or spaced; stipe scaly when young, glabrescent, densely covered in scales; lamina pinnate to 3-pinnate, often proliferous near apex; pinnae opposite to alternate, petiolulate; rachis densely covered in scales; venation free, anadromous. Sori borne medially near or at vein ending, circular, indusium peltate or rarely absent.

160–200 species throughout temperate zone and in tropical mountains.

1. Scale cover of rachis dense, often almost continuous; basal pinnule pair of lowest pinna equal to or larger than basal pinnule pair of pinna above 2
   Scale cover of rachis very sparse; basal pinnule pair of lowest pinna smaller than basal pinnule pair of pinna above .... 1. *P. zambesiacum*

2. Lamina less than 3-pinnate, no 3rd order pinnule present 3
   Lamina 3-pinnate, margin of 3rd order pinnule deeply lobed .... 2. *P. volkensii*
3. Lamina 2-pinnate; pinnule margins pungent to aristate 4
   Lamina 2-pinnate to 3-pinnatifid, gemmiferous near apex; pinnule margins shallowly crenate .... 3. *P. magnificum*
4. Pinnules moderately scaly adaxially, no gemmiferous buds 5
   Pinnules adaxially with scales on base and main vein only; 1–3 gemmiferous buds .... 4. *P. kilimanjaricum*
5. Large scales on stipe and rachis; basal pinnae up to 9 × 2 cm .... 5. *P. wilsonii*
   Large scales restricted to stipe; basal pinnae up to 14 × 3 cm .... 6. *P. transvaalense*

1. **Polystichum zambesiacum** *Schelpe* in Bol. Soc. Brot., Ser. 2, 41: 215 (1967); Roux, *Polystichum* in Africa: 291, t. 101–104 (1998). Type: Zimbabwe, Henkels Nek, Stapleford, *Schelpe* 5751 (BOL, holo.; BOL!, iso.)

Terrestrial or lithophyte; rhizome short-creeping to suberect, often branched, up to 25 mm in diameter, with linear scales up to 30 × 2 mm long, margins fringed. Fronds tufted, 5–8 per plant, arching, 100–180 cm long; stipe reddish-brown at base, paler above, up to 84 cm long, to 9 mm in diameter with adaxial groove, scales reddish-brown, large and small scales mixed, up to 16 × 10 mm; lamina firmly herbaceous to coriaceous, ± narrowly ovate, up to 89 cm long, 2-pinnate to 3-pinnatifid; rachis straw-coloured, with adaxial groove, with sparse papery pale to reddish-brown scales up to 1.8 mm long; pinnae in up to 30 free pairs, lowest pair often slightly reduced, frequently somewhat deflexed; pinnae to 27 cm long with up to 27 free pinnule pairs; basal pinnules closest to rachis simple or incised to or near to costa, acroscopic pinnules of basal pair reduced on lowest pinna, but increasing in size towards middle of lamina, pinnules in general asymmetric, narrowly ovate to narrowly triangular, up to 45 mm long, acroscopically auricled with lobate-serrate margin, obtuse; costa pale with adaxial groove at base, with few pale twisted scales up to 1.2 mm long, abaxially sparsely scaly; veins evident or obscure. Sori terminal or near terminal on abbreviated or unabbreviated vein branches, discrete at maturity, round, ± 1 mm in diameter; sporangium with 9–20 indurated annulus cells; indusium persistent, peltate circular to reniform, margin shallowly lobed to irregularly dentate, maximum radial length 0.5 mm, brown. Spores with tubercles and a reticulate pattern of ridges. Fig. 10: 11–12 & Fig. 11: 4.

TANZANIA. Lushoto District: W Usambara Mts, Shagayu Forest Reserve, NW slope of the summit, *Borhidi* et al. 84/847!; Morogoro District: Uluguru Mts, Morningside to Bondwa, July 1970, *Faden et al.* 70/316! & Lukwangule Plateau, no date, *Harris et al.* 3726!

DISTR. **T** 3, 6; Malawi, Mozambique, Zimbabwe, restricted to high mountains

HAB. Moist forest; 1800–2500 m

USES. None recorded for our area

CONSERVATION NOTES. May be vulnerable where it occurs in middle altitude forests which are heavily exploited by the local population.

2. **Polystichum volkensii** (*Hieron.*) *C.Chr.*, Index filic.: 589 (1906); Faden in U.K.W.F. ed. 2: 35 (1994); Roux, *Polystichum* in Africa: 188, t. 64–66 (1998). Type: Tanzania, Kilimanjaro, above Kibosho, *Volkens* 1520 (B!, holo.)

Terrestrial; rhizome short, erect, up to 1 cm in diameter, with persistent dense papery reddish narrowly lanceolate scales up to 15 mm long. Fronds tufted, up to 8 per plant, erect, up to 120 cm long; stipe reddish-brown at base, straw-coloured above, up to 10–52 cm long, to 1 cm in diameter, with adaxial groove, stipe bases

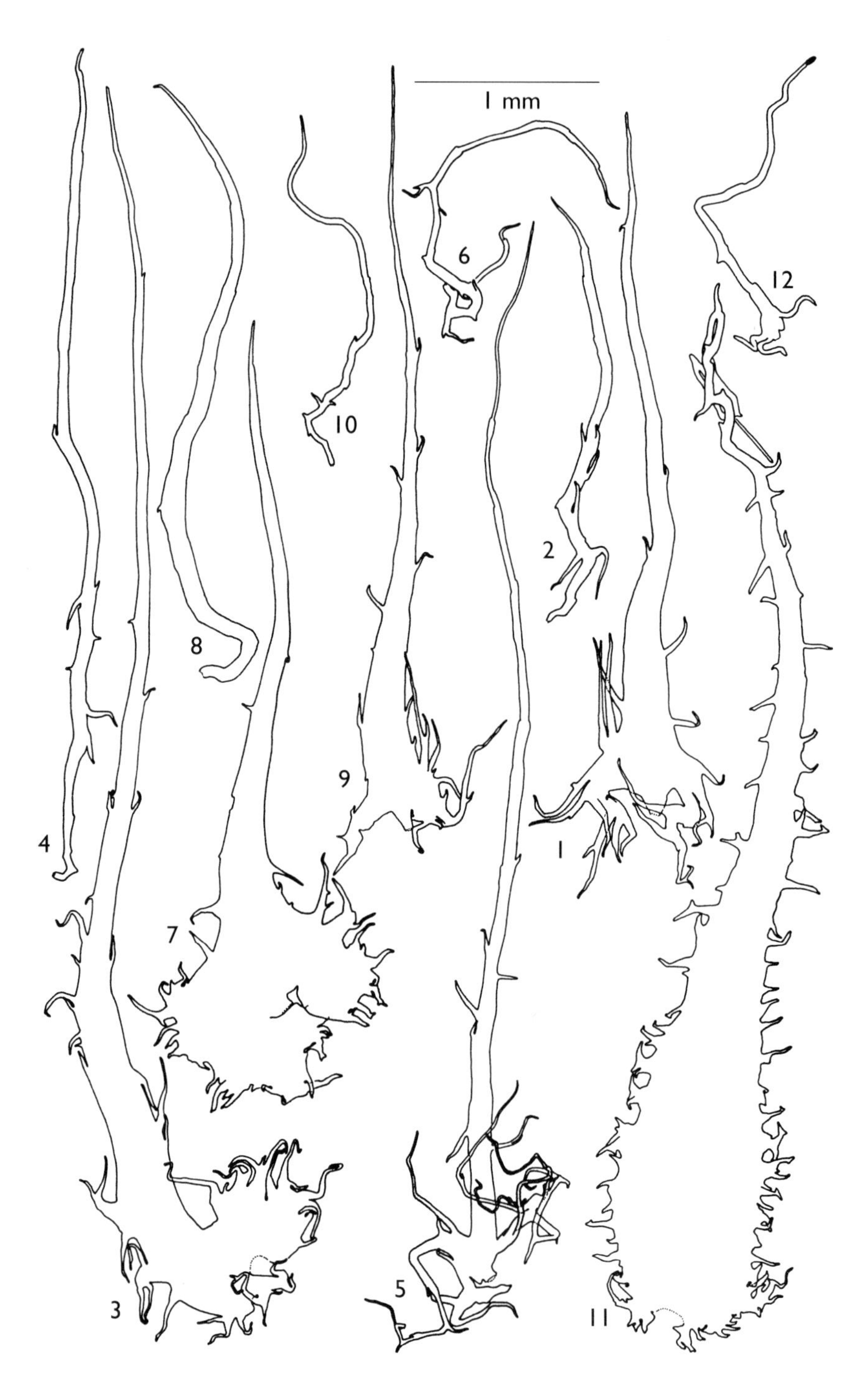

FIG. 10. *POLYSTICHUM* species, scales from rachis (uneven numbers) and lamina abaxial surface (even numbers) — **1–2,** *P. kilimanjaricum*; **3–4**, *P. magnificum*; **5–6**, *P. transvaalense*; **7–8**, *P. volkensii*; **9–10**, *P. wilsonii*; **11–12**, *P. zambesiacum*. 1–2 from *Schippers* T1456; 3–4 from *Hedberg* 965; 5–6 from *de Wilde* 5981; 7–8 from *Schippers* T1452; 9–10 from *Keay* FHI 28602; 11–12 from *Faden* 70/316. Drawn by J.P. Roux.

persistent, with dense reddish translucent shortly stalked scales, the larger elliptic to ovate, up to 34 × 10 mm, the smaller convolute, ovate; lamina herbaceous, adaxially olive green, abaxially pale green, narrowly ovate, up to 93 cm long, 3-pinnate; rachis straw-coloured with adaxial groove, with dense often folded scales up to 18 × 6 mm, smaller than, but resembling those of, the stipe, and with a single gemmiferous bud, covered in scales, near apex; pinnae oblong-attenuate, slightly falcate, basal pinnae up to 54 mm long, middle pinnae to 190 × 40 mm, lowest acroscopic pinnule slightly enlarged, a single pair of basal pinnae decrescent, often somewhat deflexed; costa straw-coloured with adaxial groove, scales dense, similar to but smaller than those on rachis; pinnules asymmetric, wedge-shaped, ovate, up to 23 × 11 mm, acroscopic margin auriculate, deeply lobed, lobes oblong, serrate to crenate, adaxially with few folded, filiform scales up to 15 mm long, abaxially covered with red to reddish-brown papery folded filiform scales up to 16.5 mm long. Sori medial on unabbreviated vein branches, mostly uniseriate, discrete at maturity, in excess of 1 mm in diameter; sporangia with 12–19 indurated annulus cells; indusium reddish to reddish-brown, peltate, circular, elliptic or amorphous in outline. Spores with reticulum of compressed ridges enclosing granulate, verruculate to echinulate surfaces. Fig. 10: 7–8 & Fig. 11: 7.

KENYA. Aberdare/Nyandarua, near Kinangop, Apr. 1922, *Fries & Fries* 2735! & W side, Aug. 1969, *Rabb & Nightingale* 7!

TANZANIA. Kilimanjaro, above Kiboscho, Oct. 1901, *Uhlig* 242! & near Mweka base hut, July 1972, *Pócs* 6718A! & above Machame route, *Schippers* T. 1452!

DISTR. **K** 3, 4; **T** 2; endemic to Aberdares/Nyandarua and Kilimanjaro

HAB. Upper moist forest zone, Hagenia zone, giant heath zone; 2800–3600 m

USES. None recorded for our area

CONSERVATION NOTES. Restricted distribution; although rare, it does not appear to be threatened.

SYN. *Aspidium volkensii* Hieron. in P.O.A. C: 86 (1895)

3. **Polystichum magnificum** *F.Ballard* in K.B. 12: 48 (1957); Faden in U.K.W.F. ed. 2: 35 (1994); Roux, *Polystichum* in Africa: 179 (1998). Type: Uganda, Mount Elgon, in the crater north of Maji ya moto, *Hedberg* 965 (K!, holo.; K!, iso.)

Terrestrial; rhizome short-creeping, branched, up to 12 mm in diameter, with dense papery reddish narrowly linear scales up to 30 × 2 mm. Fronds 6–7 per plant, erect to suberect, up to 113 cm long; stipe reddish-brown at base, straw-coloured above, up to 47 cm long, 0.9 cm in diameter, with adaxial groove, with dense reddish narrowly ovate scales up to 40 × 2 mm, margin with widely spaced projections, apex acicular, scales towards top of stipe more variable in size, larger scales narrowly ovate, up to 28 × 8 mm, smaller scales narrowly triangular and shortly stalked; lamina leathery, dark green above, slightly paler below, narrowly ovate to oblong, up to 66 cm long, 2-pinnate to 3-pinnatifid; rachis straw-coloured, up to 40 cm long, with adaxial groove, with dense reddish ovate acuminate scales and often with a single gemmiferous bud near apex, densely enclosed by reddish bud scales; pinnae oblong ovate, up to 13 cm long; costa straw-coloured with adaxial groove with dense scales, similar to but smaller than those on the rachis; pinnules inaequilaterally ovate, lobate, crenate, lowest pinnules acroscopically incised to or near to adaxial groove, adaxially and abaxially with dense reddish papery scales 4–5 mm long, simple or with short marginal outgrowths at base, often twisted; venation evident. Sori medial to inframedial, uniseriate, discrete at maturity, circular, up to 2.2 mm in diameter; sporangium with annulus of 11–17 cells; indusium persistent, peltate, incomplete to circular, erose, maximum radial length 1.14 mm, brown. Spores relatively smooth, echinulate. Fig. 10: 3–4 & Fig. 11: 6.

UGANDA. Elgon, crater north of Mayi ya moto, May 1948, *Hedberg* 965! & Suam ridge, June 1997, *Wesche* 1419!

KENYA. Elgon, Nov. 1932, *Tweedie* s.n.!
DISTR. **U** 3; **K** 3, 5; endemic to Elgon and the Bale, Arussi, and Gamu Gofa regions in southern Ethiopia
HAB. Bamboo zone or moorland, in sheltered sites; 3300–3800 m
USES. None recorded for our area
CONSERVATION NOTES. Although restricted in distribution the species is not considered to be threatened.

4. **Polystichum kilimanjaricum** *Pic.Serm.* in Webbia 27: 445 (1972); Roux, *Polystichum* in Africa: 196 (1998). Type: Tanzania, Kilimanjaro, Mandara [Bismark] Hut, *Pichi Sermolli* 5171 (FT-Pic.Serm. 20640, holo.; FT-Pic.Serm., K!, P, iso.).

Terrestrial; rhizome erect to sub-erect, up to 18 cm long, with redbrown narrowly triangular scales to 9 × 0.8 mm. Fronds tufted, 8–12 per plant, sub-erect to arching, to 100 cm long; stipe reddish-brown at base, pale above, with adaxial groove, up to 43 cm long, 8 mm in diameter, persistent base with dense brownish-red narrowly triangular scales up to 9 × 0.8 mm, often with very narrow paler brown margin with often recurved outgrowths, higher up with two types of scale, the larger ovate black-brown scales up to 18 × 7 mm, the smaller papery, straw-coloured or reddish, narrowly ovate; lamina firmly herbaceous to almost leathery, adaxially olive green, paler abaxially, oblong, up to 69 cm long, 2-pinnate; rachis straw-coloured with an adaxial groove, moderately to densely set with similar but smaller and paler scales than those on stipe, with 1–3 gemmiferous buds in the pinna axils near the apex; pinnae up to 35 free pairs, triangular to ovate, basal pinnae not or only slightly reduced in size, often somewhat deflexed, ovate, narrowly ovate or oblong attenuate, up to 14 × 3.5 cm; pinnules up to 14 free pairs, up to 20 × 10 mm; abaxially with dense acicular scales up to 2.5 mm long; costa with adaxial groove covered in acicular scales up to 7 mm long; venation evident or obscure. Sori median, uniseriate, biseriate on acroscopic auricle, discrete, circular, ± 1.4 mm in diameter; sporangium with annulus of 8–20 cells; indusium persistant, peltate, subcircular to amorphous, maximal radial length 1 mm, brown. Spores brown, ornamentation inflated or compressed tubercles or spines. Fig. 10: 1–2 & Fig. 11: 8.

TANZANIA. Mt Kilimanjaro, below 1st hut Machame route, *Schippers* TI 465! & above Mandara hut, *Schippers* T1 234A! & Kyasala river gorge above Kirisha, Feb. 1997, *Hemp* 1508!
DISTR. **T** 2; endemic to Kilimanjaro
HAB. Hagenia zone and giant heath zone, upper moist forest zone; 2600–3000 m
USES. None recorded for our area
CONSERVATION NOTES. The species is localized and rare but appears not to be threatened.

5. **Polystichum wilsonii** *H.Christ.* in Bot. Gaz. 51: 353 (1911); Roux, *Polystichum* in Africa: 149 (1998). Type: China, Szechuan, Mupin, *Wilson* 2614 (BM!, holo.)

Terrestrial or lithophytic; rhizome suberect, to 13 cm long and 1 cm in diameter, rarely branched, with dense roots and brown linear paleae to 12 × 1 mm, ending in acicular cell. Fronds tufted, 8–12 in number, to 1 m long; stipe brown near base, straw-coloured higher up, 35–45 cm long, 5 mm in diameter, with adaxial groove, sparsely to densely paleate, paleae ovate, 13–23 × 7–9 mm with fimbriate margin and darker central part; lamina herbaceous, pale to dark green, narrowly elliptic in outline, (33–)47–63 × 8–16 cm, 2-pinnate; rachis straw-coloured, densely paleate with paleae similar to but smaller than those on stipe, to 9 × 3 mm; pinnae up to 29 on each side of the rachis, the lowermost reduced and deflexed, the largest to 9 × 2 cm; pinnules up to 12 free pairs per pinna, to 12 mm long, long-aristate; venation clear, costa with dense narrowly ovate to triangular scales with long outgrowths on lower margins. Sori (sub-)terminal on abbreviated vein branches, circular, ± 1 mm in diameter; sporangium with 11–24 indurated annulus cells; indusium persistent,

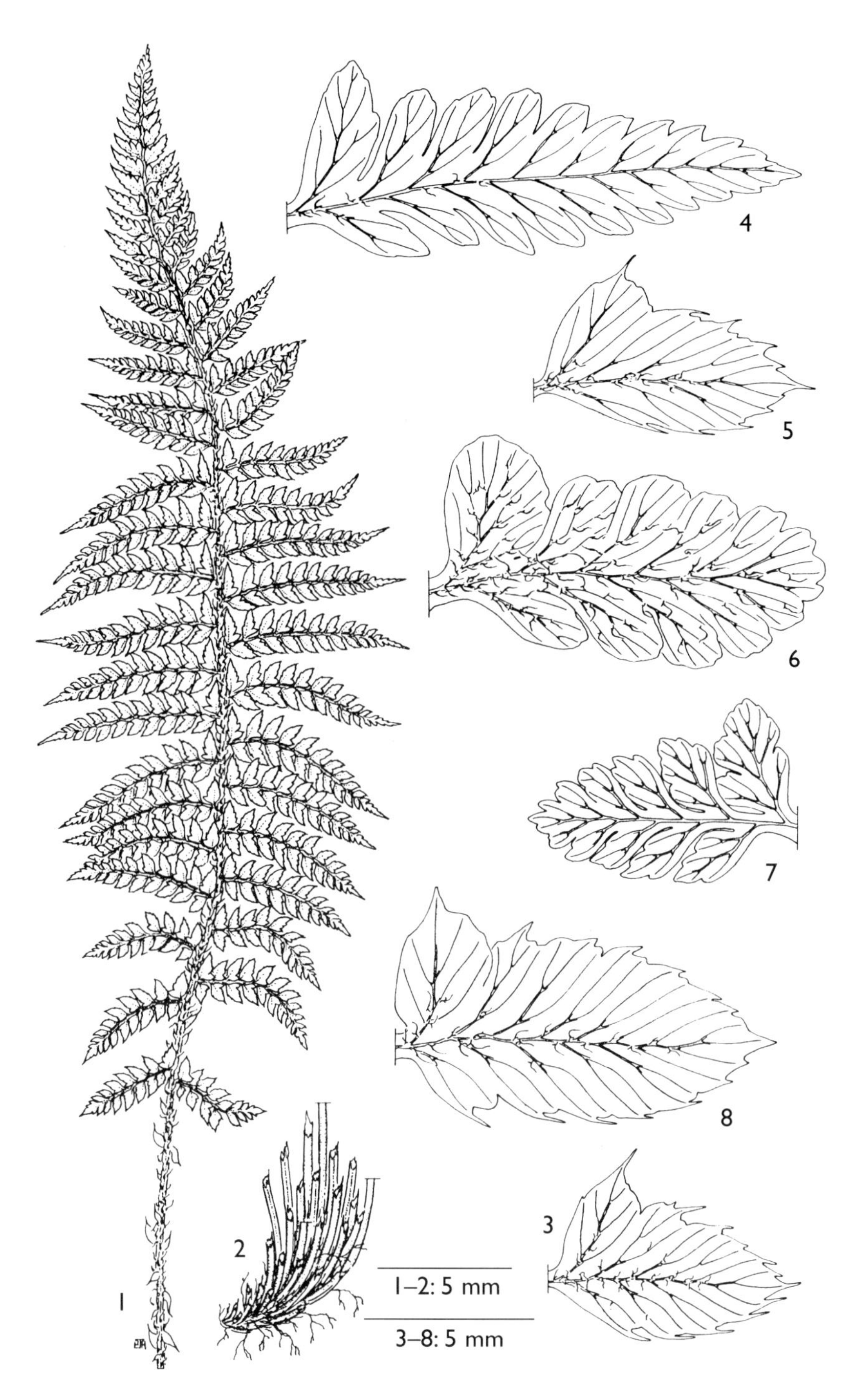

Fig. 11. *POLYSTICHUM WILSONII* — **1**, frond; **2**, rhizome; **3**, pinnule. **4–8**, pinnules of Polystichum species: **4,** *P. zambesiacum*; **5**, *D. transvaalense*; **6**, *P. magnificum*; **7**, *P. volkensii*; **8**, *P. kilimanjaricum*. 1–3 from *Gereau* 1623; 4 from *Mitchell* 135A; 5 from *Louis* 5191; 6 from *Thomerson* 550; 7 from *Schippers* T1452; 8 from *Schippers* T1456. Drawn by J.P. Roux.

straw-coloured, peltate, circular, repand to erose with small, central processes, maximum radial length 1.1 mm. Spores brown, smooth or tuberculate, spiculate. Fig. 10: 9–10 & Fig. 11: 1–3.

UGANDA. Ruwenzori, Mijusi Valley, *Hedberg* 613! & Nyamagasani valley, *Loveridge* 197!; Elgon, Benet, *Eggeling* 2454!

KENYA. Elgeyo District: Cherangani, Kamalagon [Kameligon], Aug. 1969, *Mabberley & McCall* 198!; Naivasha District: S Kinangop, Elephant Mt, June 1968, *Mwangangi* 987!; Meru District: Mt Kenya, Ithanguni, Kirui, Feb. 1970, *Faden & Evans* 70/114!

TANZANIA. Masai/Mbulu District: Ngorongoro, Empakaai, Sep. 1972, *Frame* 48!; Arusha District: slopes of little Meru, June 1968, *Vesey-Fitzgerald* 5710!; Kilimanjaro, NW Shira Plateau, Nov. 1968, *Bigger* 2282!

DISTR. **U** 2, 3; **K** 3–6; **T** 2, 3, 7; high mountains of Bioko, Cameroon, eastern Congo-Kinshasa, Ethiopia, Zimbabwe and South Africa; temperate and tropical Asia

HAB. Bamboo zone, Hagenia zone, heath zone and moorland, occasionally in moist forest, often near streams, at higher altitudes in sheltered sites; 2350–3650(?–4100) m

USES. None recorded from our area

CONSERVATION NOTES. Widespread; least concern (LC)

SYN. *P. lobatum* (Huds.) C.Presl var. *ruwensoriense* Pirotta in Savoia, Ruwenzori 1: 478 (1909). Type: Uganda, Ruwenzori, Kichuchu–Nakitava, *Roccati & Cavalli-Molinelli* s.n. (TO, holo.)

P. aculeatum (L.) Roth. var. *rubescens* Bonap., Not. Pterid. 14: 214 (1923). Type: Tanzania, Kilimanjaro, *Alluaud* 48 (P!, holo.)

P. aculeatum (L.) Roth. var. *stenophyllon* Bonap., Not. Pterid. 14: 215 (1923). Type: Kenya, Mt Kenya W, *Alluaud* 241 (P!, holo.)

P. fuscopaleaceum Alston in Bol. Soc. Brot., ser. 2, 30: 22 (1956). Type: Cameroon, Victoria District, Cameroon Mountain, SW of hut no. 2, *Keay* FHI 28 601 (BM!, holo.)

P. setiferum (Forssk.) T.Moore var. *fuscopaleaceum* (Alston) Schelpe in Bol. Soc. Brot., Ser. 2, 41: 216 (1967)

P. fuscopaleaceum Alston var. *ruwensoriense* (Pirotta) Pic.Serm. in Webbia 32: 90 (1977)

NOTE. *P. transvaalense* and *P. wilsonii* occur sympatrically and often grow side by side. *P. wilsonii* has slightly shorter and narrower fronds, more pronounced reduced and deflexed of the basal pinnae, and an indusium margin that generally is less clearly sculptured; also, the paleae occur on the rachis, while in *P. transvaalensis* they are restricted to the stipe and the lower rachis. *P. wilsonii* is prdominantly a high altitude species – although it does occur at lower altitudes as well, where it overlaps with *P. transvaalensis*. A number of intermediates occur, and a putative hybrid has been described (*P.* × *saltum*, Roux 1997).

6. **Polystichum transvaalense** *N.C.Anthony* in Contrib. Bolus Herb. 10: 146 (1982); Roux, *Polystichum* in Africa: 166 (1998). Type: South Africa, Transvaal, Pietersburg District, Woodbush Forest Reserve, *Bredenkamp & Van Vuuren* 450 (BOL!, holo.; PRE!, iso.)

Terrestrial or epiphyte; rhizome short-creeping or erect, up to 8 mm in diameter, with dense reddish-brown linear to lanceolate rhizome scales up to 14 × 1 mm, with outgrowths towards tip, apex an acicular cell. Fronds tufted, 5–17 per plant, suberect to arching; stipe reddish-brown at base, straw-coloured or green above, 20–54 cm long, 5 mm in diameter, with adaxial groove, with dense reddish-brown ovate scales 6.5–20 × 1.4–6 mm, the larger sessile, smaller stalked, tapering towards acicular cell at apex, twisted; lamina herbaceous, adaxially pale to dark green, abaxially pale green, ovate to narrowly ovate, 40–67 × 12–25 cm, 2-pinnate; rachis straw-coloured, with dense short-stalked twisted reddish brown scales up to 4.5 mm long, terminating in an acicular cell, margins an irregular fringe; pinnae to 26 free pairs, to 140 × 28 mm, the proximal slightly reduced and often deflexed, with up to 20 oblong-attenuate pairs of free pinnules, acroscopically developed towards the middle of the lamina; costa straw-coloured with adaxial groove densely scaly, scales smaller than, but resembling those of rachis; pinnules asymmetric, ovate to obliquely rhomboid, up to 15 mm long, acroscopically auriculate, basal pinnules often pinnatifid, the margins toothed to crenate, often with short awns, adaxially almost smooth or at the base of the costa with few scales, abaxially moderately covered with twisted narrowly linear to triangular scales up to 2.5 mm long; venation evident. Sori terminal or near terminal on abbreviated vein branches, discrete at maturity, circular, ± 1 mm in diameter; sporangium with 10–20 indurated annulus cells; indusium straw-coloured, peltate, round, rarely incomplete simple or often with a few central processes, 0.5–1 mm in diameter, persistent. Spores brown, ornamentation of tubercles, inflated or compressed reticulate ridges and some spines.Fig. 10: 5–6 & Fig. 11: 5.

UGANDA. Karamoja District: Mt Kadam [Debasien], Jan. 1936, *Eggeling* 2683!; Ruwenzori Mountains, Kazingo–Rwanda pass, Jan. 1932, *Hazel* 114!; Kigezi District: Luhiza, Sep. 1961, *Rose* 10311!

KENYA. Northern Frontier District: Mt Nyiru, Dec. 1972, *Cameron* 147!; Mt Elgon, Kisano Waterfall, June 1971, *Faden & Evans* 71/464!; Naivasha District: Kinangop, Brown Trout Inn, Jan. 1953, *Verdcourt* 880!

TANZANIA. Arusha District: Mt Meru, NE caldera wall, Apr. 1969, *Greenway & Fitzgerald* 13613!; Lushoto District: Shume-Magamba Forest Reserve, May 1987, *Kisena* 626!; Kigoma District: Mahali Mts, Ujamba, Aug. 1958, *Newbould & Jefford* 1731!

DISTR. **U** 1, 2; **K** 1, 3, 4, 6, 7; **T** 2–4, 6, 7; Mt Cameroon and Bioko, mountains of E Congo-Kinshasa, S Sudan, Ethiopia, Zambia, Malawi, Zimbabwe, South Africa

HAB. Moist forest, less often in dry evergreen forest or bamboo, often near water; terrestrial or occasionally epiphytic; 1350–2700 m

USES. None recorded for our area

CONSERVATION NOTES. Widespread; least concern (LC)

SYN. [*P. fuscopaleaceum* Alston var. *fuscopaleaceum* sensu Pic.Serm. in Webbia 32: 90 (1977) & in B.J.B.B. 55: 157 (1985); Faden in U.K.W.F. ed. 2: 35 (1994), *non* Alston]

NOTE. For the differences with the similar *P. wilsonii*, see the Note under that species.

10. NOTHOPERANEMA

(Tagawa) Ching in Acta Phytotax. Sin. 11: 25. (1966)

Dryopteris Adans. subgen. *Nothoperanema* Tagawa in Acta Phytotax. Geobot. 7: 199 (1938)

Medium-sized terrestrial ferns, rhizome erect or ascending, covered in dark, entire scales. Fronds in tufts. Stipes densely covered in scales, adaxial groove present, grooves of axes of higher order dwindling at base. Lamina deltoid-ovate, acuminate, tripinnatifid; lowest pinna basiscopically produced, anadromous, those above catadromous; pinna bases equal, margin entire, rarely serrate; veins dark, free, not reaching margin, apexes terminate in hydathodes. Sori round, indusiate.

A genus of five species from southern China, Taiwan, northern India, South Africa and the Madagascan region.

Nothoperanema squamiseta (*Hook.*) *Ching* in Acta Phytotaxonomica Sinica 11: 25 (1966); Faden in U.K.W.F. ed. 2: 36 (1994). Type: Bioko [Fernando Po], Clarence Peak, *Mann* 380 (K!, holo.)

Terrestrial or lithophyte; rhizome erect or short-creeping, 2–4 cm wide and deep; scales dark brown-black, to 10 × 2 mm. Fronds tufted, 6–9 per plant, erect and spreading; stipe 18–60+ cm long, 2–5 mm in diameter at base, with adaxial groove and lateral grooves, with dark brown-black linear scales 2–10 × 2 mm, patent, mixed with filiform scales of same length or smaller, leaving stubble when breaking off; in upper part of stipe fewer and smaller, linear with wider base, to 2–4 mm, frequently with glandular tip; lamina thin, pale to mid-green, red when very young, ovate to deltoid, 34 × 34 cm, 2-pinnate to 3-pinnatisect; rachis pale brown to stramineous, sparsely scaly, but with tuft at junction with pinnae; pinnae in 10–14 pairs, gradually

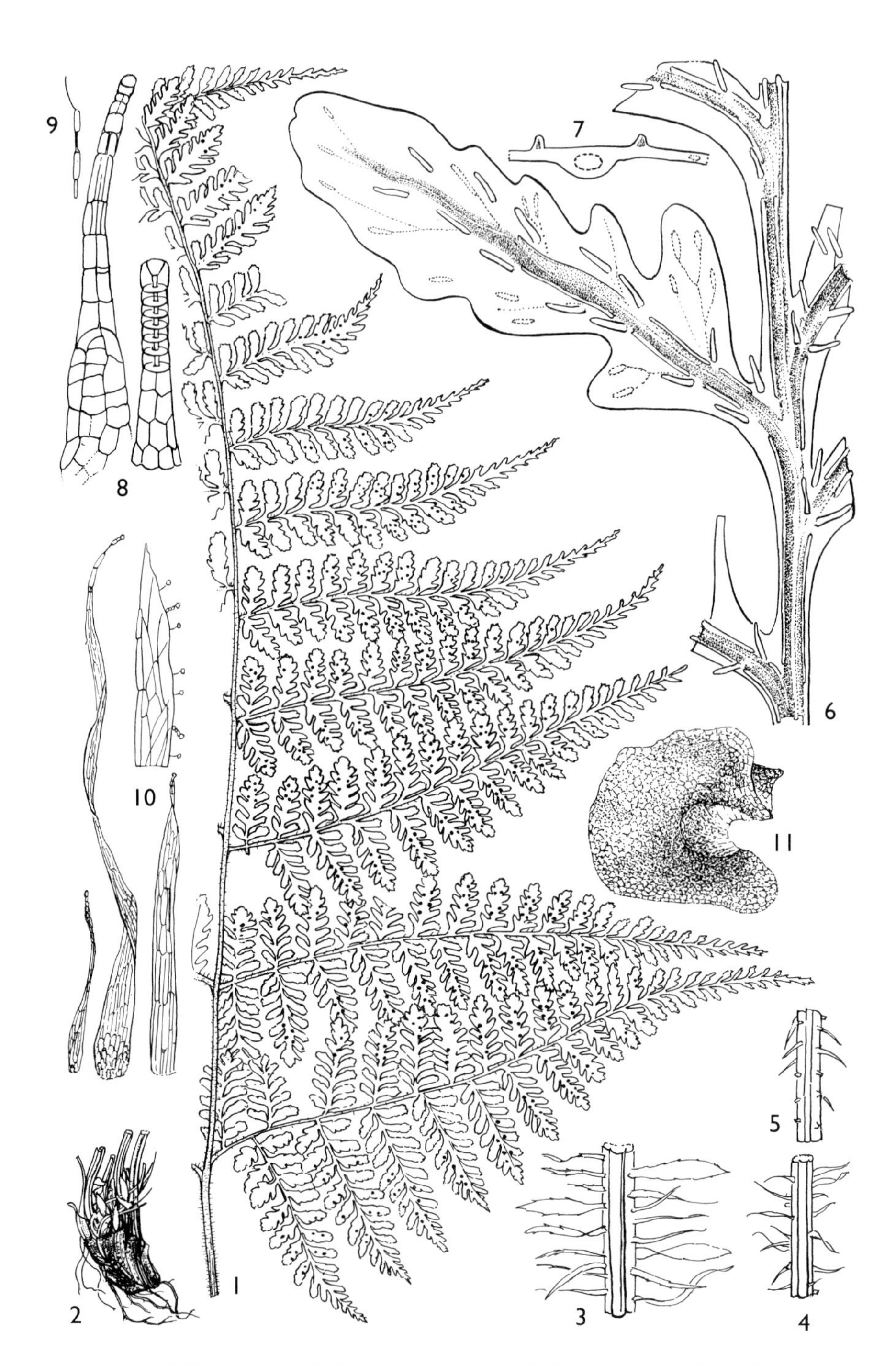

FIG. 12. *NOTHOPERANEMA SQUAMISETA* — **1**, frond, × 0.5 ; **2**, rhizome, × 0.5; **3–5**, stipe near base/mid/apex, × 2 ; **6**, pinnule, × 7; **7**, pinnule transverse section, × 27; **8–9**, trichomes from adaxial surface, × 70; **10**, scales, × 8 (margin × 50); **11**, indusium, × 25. 1–5 from *Faden et al.* 71/862; 6–9 from *Hall* s.n.; 10 from *Lock* 69/1; 11 from *Grimshaw* 93/865. Drawn by Monika Shaffer-Fehre.

merging into pinnatifid tip marked by decurrent lamina; pinnules in 8–12 pairs, opposite or sub-opposite, lower ones with lobed or crenate margin, basal acroscopic segment always more deeply cut or extended along costa; mid-vein in segment with ± 2 pairs of veinlets, not reaching margin; adaxial lamina glabrous, veins and veinlets with few relatively large multiseptate trichomes (0.5)–1 mm long. Sori frequently constrained to top half, rarely present throughout pinnule, 1–1.4 mm in diameter; indusium 0.8–1 mm in diameter, with a plain, heavily undulating margin, deep notch to its centre aligned with the underlying veinlet, backward-pointing to the next larger vein or parallel to costa, stalk 0.3 mm long, with thick, knobbly radial walls. Spores monolete, oval, 0.05 mm long, 0.03 mm wide, with widely spaced, coarse tubercles. Fig. 12.

UGANDA. Ruwenzori, 6.5 km N of Kilembe, Feb. 1969, *Lock* 69/1!
KENYA. Naivasha District: S Kinangop, Nov. 1957, *Molesworth Allen* 3638!; Mt Kenya, lower end of Teleki Valley, Dec. 1957, *Verdcourt* 2050!; Meru District: Marimba Forest on Ithanguni, Feb. 1970, *Faden* 70/76!
TANZANIA. Kilimanjaro, above Mandara hut, Oct. 1993, *Grimshaw* 93/865!; Morogoro District: N Uluguru Forest Reserve, Lupanga Peak W, 1981, *Hall* s.n.; Iringa District: Mufindi, Kigogo Forest, Dec. 1961, *Vesey-Fitzgerald* 3655!
DISTR. **U** 2; **K** 3, 4; **T** 2, 6, 7; from Bioko and Cameroon to Ethiopia and south to South Africa; Madagascar, Asia
HAB. Moist forest, often near water, or bamboo forest; 1850–2950 m
USES. None recorded for our area
CONSERVATION NOTES. Widespread; least concern (LC)

SYN. *Nephrodium squamisetum* Hook., Sp. Fil. 4: 140 t.268 (1862)
Aspidium squamisetum (Hook.) Kuhn, Fil. Deck.: 24 (1867); [Fil. Afr. 24 (1868)]
Nephrodium buchananii Baker, Syn.: 498 (1874). Type: South Africa, Natal, *Buchanan* 108 (K!, holo)
Lastrea buchananii (Baker) Bedd., Handb.: 255 (1883)
Dryopteris buchananii (Baker) Kuntze, Rev. Gen. Pl. 2: 812 (1891); Sim: 108 (1915)
Dryopteris squamiseta (Hook.) Kuntze, Rev. Gen. Pl. 2: 813 (1891)

NOTE. Ching states that his new genus occupies an intermediate systematic position between *Dryopteris*, *Peranema* and *Diacalpe*. The genus agrees with *Peranema* D.Don and *Diacalpe* Blume in habit, leaf texture and colour, pattern of pinnation and particularly in the presence on costa and costules above of peculiar thick, stout, rufo-brown setae (trichomes) at the insertion of veins and veinlets, but differs above all in the dryopteroid indusium, the position of which is superior in relation to the sorus.

11. DRYOPTERIS

Adans., Fam. Pl. 2: 20, 551 (1763); Kramer in Kubitzki *et al.*, Fam. Gen. Vasc. Pl. 1: 110–112 (1990)

Plants terrestrial or epilithic. Rhizome erect to short-decumbent, simple or sparsely branched, densely scaled. Fronds closely spaced, axes shallowly sulcate, sulci confluent or not; lamina anadromous, catadromous towards apex, herbaceous to firmly herbaceous, 2–4-pinnate-pinnatifid, basal pinna pair mostly basiscopically developed, adaxially proliferous along rachis or not; venation free, lateral veins in lobes simple, forked, or pinnately branched, vein branches end in teeth near margin, endings often slightly enlarged; indumentum composed of unicellular glands, uniseriate pluricellular, isocytic (all cells similar) or moniliform hairs, and scales occurring on rhizome, frond axes and lamina. Sori circular, medial to inframedial on predominantly anadromous (often modified) vein branches; sporangium stalk simple, glandular and/or haired; sori exindusiate or indusiate, indusium firmly herbaceous, deltoid to reniform, entire to erose, often glandular along margin and on surface. Spores ellipsoidal, monolete, pale to dark brown, perispore folded to form reticulate ridges or echinulate.

A genus of about 150 species with a nearly world-wide distribution.

1. Sori exindusiate 2
   Sori indusiate 6
2. Lamina proliferous, mostly with one or more buds adaxially on the rachis near the lamina apex ..... 7. *D. manniana*
   Lamina never proliferous 3
3. Scales abaxially along the costae and costules often bullate .... 3. *D. fadenii*
   Scales abaxially along the costae and costules never bullate 4
4. Lamina axes and veins closely set with mostly 3-celled isocytic hairs .... 12. *D. tricellularis*
   Lamina axes and veins mostly with scales, if hairs present then not predominantly 3-celled 5
5. Stipe scales up to 18 × 6 mm, linear-acuminate to narrowly lanceolate, irregularly denticulate, with glands and long outgrowths ; **K** 4, 5; **T** 6 .... 4. *D. filipaleata*
   Stipe scales up to 12 × 2.5 mm, narrowly ovate to lanceolate, cordate to cuneate, with glands ; **U** 2 .. 10. *D. ruwenzoriensis*
6. Scales abaxially along the costae and costules bullate 7
   Scales abaxially along the costae and costules never bullate 8
7. Lamina to 2-pinnate-pinnatifid; pinna rachis winged; indusium small and deltoid .... 3. *D. fadenii*
   Lamina to 4-pinnate-pinnatifid; pinna rachis not winged; indusium reniform, entire or glandular along the margin (rarely also on the indusium surface) .... 5. *D. kilemensis*
8. Sori at or near apex of mostly shortened vein branches 1. *D. antarctica*
   Sori medial to inframedial on unmodified vein branches 9
9. Pinnae mostly set at about 45° or less to the rachis; segments narrowly trullate, narrowly triangular, narrowly rhomboid, or oblong-obtuse, obtusely denticulate to crenate .... 2. *D. athamantica*
   Pinnae mostly set at 45° or more to the rachis; segments ovate, oblong-obtuse, or oblong-acute, dentate or serrate 10
10. Lamina axes and veins variously set with rigid isocytic hairs 11
    Lamina axes and veins variously set with hairs and scales, but never rigid isocytic hairs 12
11. Lamina axes and veins closely set with short, mostly 3-celled isocytic hairs .... 12. *D. tricellularis*
    Lamina axes and veins variously set with pluricellular isocytic hairs .... 11. *D. schimperiana*
12. Costae and veins abaxially eglandular; lamina hairs never glandular near base; perispore tuberculate .. 9. *D. rodolfii*
    Costae and veins abaxially glandular or eglandular; lamina hairs often glandular near base; perispore folded to form reticulate ridges 13
14 Lamina axes and veins mostly with oblong glands 60–260 μm long, and 2-celled hairs; if eglandular then stomata 34–72 μm long .... 8. *D. pentheri*
   Lamina axes and veins with costly with clavate glands 54–164 μm long, 2-celled hairs absent; if eglandular then stomata 30–62 μm long .... 6. *D. lewalleana*

1. **Dryopteris antarctica** (*Baker*) *C.Chr.*, Ind. filic., Suppl. 1: 29 (1913); Fraser-Jenkins in Bull. Brit. Mus. Nat. Hist., Bot. 14: 195, 208 (1986); Faden in U.K.W.F. ed. 2: 36 (1994); J.P. Roux, Conspect. southern Afr. Pterid.: 1127 (2001). Type: South Indian Ocean, Amsterdam Island, *G. Staunton* s.n. (BM!, holo.)

Terrestrial or epilithic; rhizome erect to suberect, short, closely branched, up to 5 mm in diameter, with dense ferrugineous to castaneous lanceolate to narrowly ovate scales up to 9 × 3 mm, often bicolorous, entire or with scattered glands along margin. Fronds tufted, arching, to 77 cm long; stipe castaneous near base, stramineous higher up, to 40 cm long, to 4.5 mm in diameter, moderately set with glands and spreading scales to 11 × 4.5 mm, similar to those on rhizome; lamina herbaceous to thinly herbaceous, ovate, to 45 cm long, to 3-pinnate; rachis greenish, narrowly winged towards apex, often set with glands and with moderate to sparse scales similar to those on stipe, up to 4 × 2 mm; pinna in up to 17 stalked pairs, basal pinnae inaequilaterally ovate, those higher up symmetric, narrowly ovate to oblong-acuminate, basal pinna pair mostly longest, basiscopically developed, up to 85 × 15 cm, with up to 8 stalked pinnule pairs; pinna-rachis narrowly winged for most of its length, glandular, and sparsely scaled; pinnules petiolate, narrowly ovate to oblong-obtuse, 1-pinnate, often acroscopically developed, acroscopic pinnule on basal pinnae up to 28 × 12 mm, basiscopic pinnule on basal pinnae up to 65 × 25 mm; ultimate segments broadly ovate to oblong-obtuse, basiscopically decurrent, up to 18 × 9 mm, with strongly dentate lobes, adaxially glabrous or with sparse glands along veins, often also with hairs and filiform scales near base, abaxially sparsely to moderately set with glands, often also with hairs, and with sparse scales similar to, but smaller than those on pinna-rachis. Sori at or near apex of mostly shortened anadromous vein branch, discrete at maturity, circular, up to 1.2 mm in diameter; sporangium stalk simple; indusium brown, firmly herbaceous, reniform, entire to erose, often glandular along margin, up to 1.2 mm in diameter. Spores with compressed reticulate folds, densely echinulate, 44–58 × 32–42 mm.

UGANDA. Kigezi District: Mt Mgahinga summit, 24 Aug. 1938, *A.S. Thomas* 2460!
KENYA. South Nyeri District: Aberdares National Park, above Gura Falls, 9 Feb. 1969, *Faden* 69/156!; Meru District: Kirui on the slopes of Ithanguni, 28 Feb. 1970, *Faden* 70/105! & Kiandongoro, track above Tucha, 24 Oct. 1971, *Faden & Faden* 71/883!
TANZANIA. Kilimanjaro, 7 Mar. 1934, *Schlieben* 4898!
DISTR. **U** 2; **K** 3/4, 4, **T** 2; scattered along the higher East and Central African mountains to the Western Cape mountains in South Africa; Réunion and Amsterdam Is. in the South Atlantic
HAB. Moorlands, montane grasslands, giant heath zone and bamboo zone; 2500–3200 m
USES. None recorded for our area
CONSERVATION NOTES. Widespread; least concern (LC)

SYN. *Nephrodium antarcticum* Baker in J.L.S. 14: 479, 480 (1875); Faden in U.K.W.F., ed. 2: 36, t. 175 (1994)
Dryopteris callolepis C.Chr. in N.B.G.B. 9: 177 (1924); Schelpe, F.Z., Pterid.: 223 (1970); Jacobsen, Ferns S. Afr.: 439, fig. 330, map (1983); Pic.Serm. in B.J.B.B. 55: 157 (1985); Schelpe & N.C.Anthony, F.S.A., Pterid.: 249, t. 85, map 217 (1986); J.E.Burrows, S. Afr. Ferns: 304, t. 51/5, fig. 71/309, map (1990); Kornaś & K.A.Nowak in Acta Soc. Bot. Poloniae 61: 160 (1992); Kornaś et al. in Prace Bot. 25: 30, map 48b (1993). Type: Kenya, Aberdares, *R.E. & T.C.E. Fries* 2554 (BM!, holo.)

NOTE. Widén *et al.* in Acta Bot. Fennica 164: 1–56 (1999) tentatively separated *Dryopteris callolepis* from *D. antarctica* for 'geograpical convenience' and because of their distinct biochemistry. Detailed morphological studies on material from Africa, St Paul and Réunion, however, suggest them to be conspecific. I therefor have no hesitation in recognising a single species, *D. antarctica.* Vida, in Helv. Chim. Acta 56: 2129 (1973) report the species as tetraploid: 2n = 164 ±4.

2. **Dryopteris athamantica** (*Kunze*) *Kuntze*, Rev. Gen. Pl. 2: 812 (1891); C.Chr., Ind. Filic.: 253 (1905); Tardieu-Blot in Mém. I.F.A.N. 28: 151, t. 29/7 (1953); Alston, Ferns W.T.A.: 70 (1959); Tardieu-Blot, Fl. Cameroon, Pterid.: 262, t. 40/7 (1964); Schelpe, F.Z., Pterid: 221 (1970); Schelpe, Expl. Hydrobiol. L. Bangweolo & Luapula: 89 (1973); Schelpe & Diniz, Fl. Moçamb., Pterid.: 237 (1979); Kornaś, District: Ecol. Pterid. Zambia: 107, map 71a (1979); Jacobsen, Ferns S. Afr.: 433, fig. 326, map 159 (1983); Pichi Sermolli in B.J.B.B. 55: 157 (1985); Schelpe & Anthony, F.S.A., Pterid.: 247, t. 84/2, map 213 (1986); Burrows, S. Afr. Ferns.: 300, t. 50/4, fig. 70/305, map (1990); Faden in U.K.W.F., ed. 2: 36, t. 175 (1994); J.P. Roux, Conspect. southern Afr. Pterid.: 127 (2001). Type: South Africa, ad portum Natalensem, Feb. 1842, *Gueinzius* s.n. (LZ†, holo.; K!, lecto., here designated)

Terrestrial or epilithic; rhizome short-decumbent to suberect, rarely branched, to 13 mm in diameter, with ferrugineous oblong-acuminate to narrowly lanceolate scales up to 18 × 2 mm, entire. Fronds tufted, erect, up to 1.3 m long; stipe stout, proximally castaneous, stramineous higher up, up to 54 cm long and 14 mm in diameter, with ferrugineous linear to oblong scales up to 24 × 2.5 mm, acuminate, entire, with long twisted filiform outgrowths or with a few scattered glandular cells; lamina firmly herbaceous, narrowly ovate, up to 78 cm long, up to 2-pinnate-pinnatifid; rachis stramineous to greenish, narrowly winged towards apex, moderately to sparsely set with twisted ferrugineous to stramineous linear to filiform scales up to 6 × 0.6 mm and hairs; pinnae up to 23 stalked pairs, basal pinnae inaequilaterally broadly ovate, those higher up symmetrically lanceolate, basal pinna pair mostly longest, mostly basiscopically developed, up to 19 × 12 cm, with up to 8 stalked pinnule pairs; pinna-rachis narrowly winged for most of its length, with sparse scales up to 4 × 0.4 mm, similar to, but smaller than those on rachis; pinnules lanceolate, acroscopically developed, acroscopic pinnule on basal pinnae up to 48 × 28 mm, basiscopic pinnule on basal pinnae up to 72 × 30 mm; costa narrowly winged; segments narrowly trullate, narrowly triangular, narrowly rhomboid or oblong-obtuse, up to 20 × 7 mm, basiscopically decurrent, often shallowly lobed, lobes denticulate to crenate, adaxially glabrous or with a few filiform scales along the costae, abaxially sparsely set with filiform scales, uniseriate hairs bearing one or more glandular cells near base, isocytic hairs, and often also with unicellular glands (50–)78(–142) mm long. Sori inframedial on the predominantly anadromous vein branches, discrete or touching at maturity, circular, up to 1.6 mm in diameter at maturity; sporangium stalk simple or haired; indusium brown, firmly herbaceous, reniform, entire to erose, up to 1.7 mm in diameter. Spores ellipsoidal monolete, perispore with narrow to broad reticulate ridges, granulate, 38–52 × 26–36 mm.

UGANDA. West Nile District: Kango, Apr. 1941, *Eggeling* 4264!; Ankole District: Nyagoma-Rugongo, Buhweju, 1 Aug. 1986, *Rwaburindore* 2285!; Mbale District: Mt Elgon, *E. James* s.n.!
KENYA. Trans-Nzoia District: Kiptorok, Kitale, Mlimani area, 14 June 1971, *Faden* 71/476!; North Kavirondo District: Kakamega forest, 27 Nov. 1969, *Faden & Evans* 69/2054!
TANZANIA. Ngara District: Keza, Bushubi, 20 March 1961, *Tanner* 5901!; Kigoma District: Kasye Forest, 26 March 1994, *Bidgood et al.* 2993!; Songea District: Matengo Hills, 6 March 1956, *Milne-Redhead & Taylor* 9040!
DISTR. **U** 1–5; **K** 3, 5; **T** 1, 4, 7, 8; throughout West and southern Africa
HAB. Grassland, scattered tree grassland, miombo woodland, often near water or in shade of rocks; (1000–)1200–2000(–3050) m
USES. Against intestinal parasites (fide *Tanner*)
CONSERVATION NOTES. Widespread; least concern (LC)

SYN. *Aspidium athamanticum* Kunze in Linnaea 18: 123 (1844); Hieron. in P.O.A. C: 86 (1895)
Lastrea athamantica (Kunze) T.Moore in J. Bot. 5: 311 (1853); Pappe & Rawson, Syn. Fil. Afr. Austr.: 13 (1858)
Nephrodium athamanticum (Kunze) Hook., Sp. fil. 4: 125, t. 258 (1862); Hook. & Baker, Syn. Fil.: 277 (1874); Sim, Ferns S. Afr.: 183, t. 102 (1892)
Lastrea plantii T.Moore in J. Bot. 5: 227 (1853). Type: South Africa Natal, Mooi River, *Plant* 313 (BM!, holo.)

Nephrodium eurylepium A.Peter in F.R., Beih. 40, 1: 57 et App.: 3 (1929). Type: Tanzania, Buha [Uha] District, *Peter* 37743 (K!, syn.); Tanzania, Kigoma District, Ujiji, *Peter* 37290 (K!, syn.)

NOTE. Vida in Helv. Chim. Acta 56: 2129 (1973) report the species as diploid: 2n = 80 ±2.

3. **Dryopteris fadenii** *Pic.Serm.* in Webbia 37: 333, 336, fig. 3 & 4 (1984); Faden in U.K.W.F., ed. 2: 36, t. 175 (1994). Type: Kenya, Naivasha/Kiambu District, Sasumua Dam, *Faden, Evans & Cameron* 71/68 (FT-Pic.Serm., holo.; BOL!, K!, Herb. Reichstein, Basel, iso.)

Terrestrial; rhizome short-decumbent, up to 16 mm in diameter, with crowded stipe bases and brown to dull ferrugineous scales, larger scales up to 31 × 5 mm, smaller scales linear-acuminate to ovate, irregularly set with short bifid teeth, glands, and irregular outgrowths. Fronds 5–7 per plant, tufted, erect to arching, up to 1.5 m long; stipe proximally castaneous, stramineous higher up, up to 61 cm long and 10 mm in diameter, near base with dense scales up to 24 × 10 mm, those higher up fugaceous; lamina firmly herbaceous, ovate to broadly ovate, up to 92 × 62 cm, 2-pinnate to 2-pinnate-pinnatifid; rachis greenish to stramineous, narrowly winged towards apex, initially with sparse to moderate stramineous to ferrugineous, scales up to 6 × 1.5 mm; pinnae up to 8 stalked pairs, subsessile and adnate towards apex, basal pinnae triangular, lanceolate, or oblong-acuminate towards lamina apex, basal pair longest, basiscopically developed, up to 32 × 14 cm; pinna-rachis narrowly winged for most of its length, subglabrous adaxially, initially moderately set with scales up to 5 × 2.5 mm on abaxial surface; pinnules petiolate, narrowly lanceolate to oblong-acuminate, often slightly basiscopically developed, acroscopic pinnule on basal pinnae up to 86 × 25 mm, basiscopic pinnule on basal pinnae up to 106 × 38 mm; segments sessile, lanceolate to oblong, basiscopically decurrent, up to 18 × 8 mm, shallowly lobed to shallowly dentate, adaxially glabrous or with a few scattered hairs along costule, abaxially sparsely to moderately set with moniliform or isocytic hairs and filiform to subulate scales. Sori medial to inframedial, discrete or touching at maturity, circular, up to 2 mm in diameter; sporangium stalk simple, glandular, or with a short, few-celled hair; indusium absent or present, brown, firmly herbaceous, cordate to reniform, repand to erose, margin glandular or eglandular, up to 1 mm in diameter. Spores variously set with prominent tubercles and/or ridges, ruminate, 30–78 × 20–50 mm. Fig. 13: 1.

UGANDA. Ruwenzori Mountains, Mobuku Valley, *Esterhuysen* 25178a! & Ruwenzori, Mijusi Valley, *Hedberg* 614!; Masaka District: Bigo, Bujuku River, 23 July 1952, *Osmaston* 1687!
KENYA. Naivasha District: Sasamua Dam, ± 1.5 km downstream along river, 24 Jan. 1971, *Faden* 71/75!; Meru District: Marimba, Ithangune forest at the base of Kirue, *Faden et al.* 69/757B!; Kericho District: Kimugung River, ± 5 km NW of Kericho, 10 June 1972, *Faden et al.* 72/285!
TANZANIA. Arusha District: Mt Meru, W slopes of above Olkakola estate, 31 Oct. 1948, *Hedberg* 2402!; Iringa District: Mufindi, Lugodo Factory, 14 Aug. 1971, *Perdue & Kibuwa* 11111!; Rungwe District: Rungwe Mt, ± 4 km SE of Isongole and 2 km WSW of Shiwaga Crater, 10 June 1992, *Mwasumbi* 16228!
DISTR. **U** 2, 4; **K** 3–5; **T** 2, 3, 6, 7; confined to East Africa
HAB. Wet montane forests, riverine forests, streambanks, *Albizia-Podocarpus* forests, bamboo-*Podocarpus* forests, giant heath zone; 1700–2500(–3500) m
USES. None recorded for our area
CONSERVATION NOTES. Widespread; least concern (LC)

NOTE. Vida, in Helv. Chim. Acta 56: 2130 (1973) report the species under the name *D. pentheri* as diploid: 2n = 82 ±4.

4. **Dryopteris filipaleata** *J.P.Roux* in Bothalia 34: 28, fig. 1 & 2 (2004). Type: Tanzania, Morogoro District, Uluguru mountains, Mwere valley, *Faden et al.* 70/596 [BOL!, holo. (2 sheets); K!, iso. (2 sheets)]

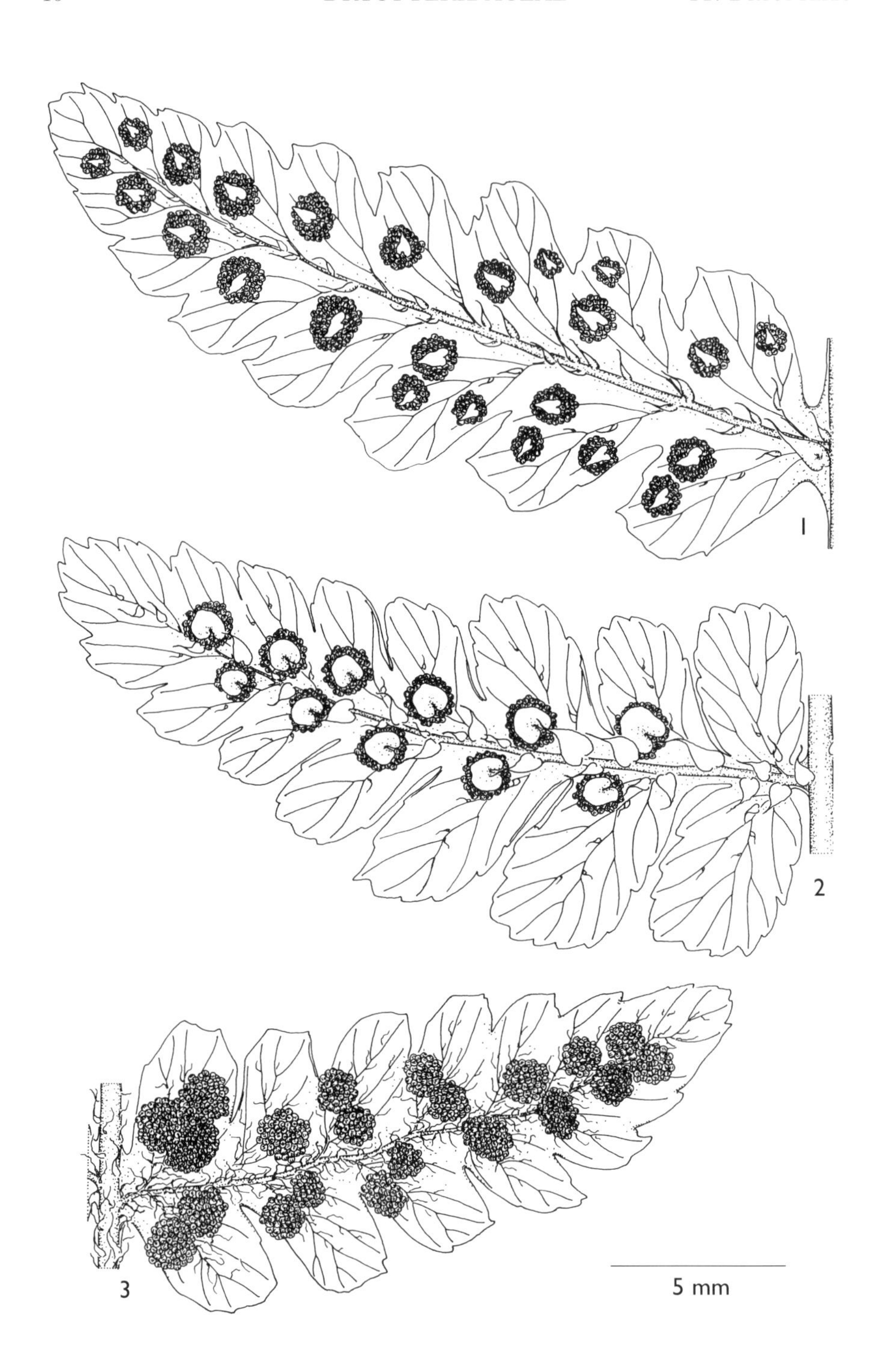

FIG. 13. — **1**, *DRYOPTERIS FADENII*, abaxial view of fertile pinnule; **2**, *D. KILEMENSIS*, abaxial view of fertile pinnule; **3**, *D. MANNIANA*, abaxial view of fertile pinnule. 1 from *Faden et al.* 71/68; 2 from *MacLeay* 93; 3 from *Van Someren* 438. Drawn by J.P. Roux.

Terrestrial; rhizome short-decumbent, up to 12 mm in diameter, with closely set stipe bases and ferrugineous to castaneous linear to narrowly lanceolate scales up to 15 × 5 mm, acuminate, irregularly set with scattered capitate glands and long outgrowths. Fronds tufted, suberect to arching, up to 1.1 m long; stipe proximally castaneous, stramineous higher up, up to 61 cm long and 7 mm in diameter, with proximally dense and distally sparse ferrugineous scales up to 18 × 6 mm; lamina ovate to broadly ovate, up to 59 × 43 cm, 2-pinnate-pinnatifid to 3-pinnate; rachis stramineous, initially densely scaly, glabrescent, scales ferrugineous to castaneous, linear to filiform, up to 4 × 0.5 mm, irregularly denticulate; pinnae up to 13 stalked pairs, increasingly more broadly attached and basiscopically decurrent along rachis towards lamina apex, basal pair longest and mostly conspicuously basiscopically developed, inaequilaterally triangular, ovate, oblong-acuminate to lanceolate, up to 27 × 13.5 cm, those higher up mostly near symmetrical, with up to 7 pinna pairs; pinna-rachis narrowly winged towards apex, with some linear to filiform scales up to 3 × 0.3 mm; pinnules petiolate, increasingly more broadly attached and basiscopically decurrent along pinna-rachis towards pinna apex, up to 78 × 26 mm, proximal basiscopic pinnules slightly basiscopically developed, narrowly lanceolate to oblong-acuminate; costa narrowly winged, with sparse scales up to 2.2 × 0.1 mm; segments and lobes up to 15 × 7 mm, ovate-obtuse to oblong-obtuse, basiscopically decurrent, shallowly lobed to denticulate, adaxially glabrous or with few hairs and filiform scales along costa, abaxially sparsely set with moniliform hairs up to 0.6 mm long on and between veins. Sori predominantly 2-seriate along pinnules, 2-seriate on lobes in larger plants, medial to inframedial on predominantly anadromous vein branches; exindusiate; sporangium stalk simple, glandular or haired. Spores with low reticulate ridged and bulges, 32–54 × 18–34 μm.

KENYA. South Nyeri District: Thiba Fishing Camp, 31 July 1977, *Gilbert & Rankin* 4821!; Meru District: Meru, upper forest, Aug. 1949, *H.D. van Someren* 493!; Kisumu-Londiani District: Kisumu, Feb. 1915, *Dummer* 1524!, 1727! & 1728!

TANZANIA. Morogoro District, Uluguru mountains, Morningside to Bondwa, July 1970, *Faden et al.* 70/351! & Northern Nguru, Kanga mountain, 2 Dec. 1987, *Lovett & Thomas* 2800! & "Bagamoyo District, mainland west of Zanzibar", March 1885, *Last* s.n.!

DISTR. **K** 4, 5; **T** 6; restricted to the mountainous areas of tropical East Africa

HAB. Moist evergreen forests, often along streambanks; 1250—2000 m

USES. None recorded for our area

CONSERVATION NOTES. Restricted in distribution, but appears not to be threatened.

5. **Dryopteris kilemensis** (*Kuhn*) *Kuntze*, Rev. Gen. Pl. 2: 813 (1891), as *kilmensis*; Schelpe, F.Z., Pterid.: 222 (1970) & Expl. Hydrobiol. Bassin L. Bangweolo & Luapula, 8(3), Ptérid.: 90 (1973); Schelpe & Diniz, Fl. Moçamb., Pterid: 239 (1979); Jacobsen, Ferns S. Afr.: 438, fig. 329, map 161 (1983); Pichi Sermolli in B.J.B.B. 55: 157 (1985); Burrows, S. Afr. Ferns: 303, t. 51/4, fig. 71/308, map (1990); Faden in U.K.W.F. ed. 2: 36, t. 175 (1994); J.P. Roux, Conspect. southern Afr. Pterid.: 124 (2001). Type: Tanzania, Kilimanjaro, Chagga, *Kersten* 46 (B, holo.)

Terrestrial or epilithic; rhizome erect to suberect, short, up to 15 mm in diameter, with crowded stipe bases and ferrugineous subulate to lanceolate scales up to 30 × 4 mm, entire or with a few filiform outgrowths near apex. Fronds tufted, arching, up to 1.5 m long; stipe proximally castaneous, stramineous higher up, up to 82 cm long and 9 mm in diameter, proximally densely scaly, subglabrous higher up, scales ferrugineous to castaneous, filiform to broadly ovate, up to 40 × 9 mm, entire or with a few filiform outgrowths near base or apex, occasionally with a few scattered glands and short uniseriate hairs; lamina herbaceous, broadly ovate, up to 63 cm long, up to 4-pinnate-pinnatifid; rachis stramineous, becoming narrowly winged towards apex, with sparse scales up to 6 × 3 mm, similar to, but smaller than those on stipe; pinnae with up to 14 stalked pairs, basal pair inaequilaterally ovate, ovate to lanceolate towards lamina apex, basal pair longest, basiscopically developed, up to

35 × 23 cm, with up to 12 stalked pinnule pairs; pinna-rachis often closely set with unicellular glandular hairs and scales, abaxially with sparse scales up to 4.5 × 2.7 mm, similar to, but smaller than those on rachis; pinnules ovate, acroscopic pinnule on basal pinnae up to 105 × 29 mm, basiscopic pinnule on basal pinnae up to 150 × 70 mm, with up to 9 stalked segment pairs; costa with sparse scales similar to, but smaller than those on pinna-rachis; segments petiolate, ovate, with up to 3 stalked ultimate segment pairs, spaced, basiscopic segment on the basal pinnule to 44 × 20 mm, acroscopic segment on basal pinnule up to 28 × 14 mm; ultimate segments oblong-obtuse, deeply lobed, lobes serrate to obtusely dentate, glabrous adaxially, abaxially glabrous or often with glands along veins, with isocytic hairs, and scales to 1.5 × 0.7 mm. Sori circular, inframedial, discrete at maturity, to 1 mm in diameter; sporangium stalk simple, glandular, or rarely with a pluricellular uniseriate hair; indusium brown, firmly herbaceous, reniform, entire or glandular along margin (rarely also on surface), up to 1 mm in diameter. Spores with low tubercles and/or long reticulate ridges, minutely rugulose to minutely scabrous, 32–42 × 20–26 mm. Fig. 13: 2, p. 40.

UGANDA. Ruwenzori, Mobuku valley, 7 Jan. 1939, *Loveridge* 330!; Ankole District: W Ankole, Buhweju, Isingiro, 2 Nov. 1992, *Rwaburindore* 3492!; Mt Elgon, Sasa trail, 27 Dec. 1996, *Wesche* 565!

KENYA. Mt Kenya, Castle Forest Station, 17 Oct. 1971, *Faden et al.* 71/871!; Embu District: below Castle forest station, 19 Dec. 1972, *Gillett & Holttum* 20088!; Teita District: Vuria Hill, 8 May 1985, *Faden et al.* 164!

TANZANIA. Moshi District: Kilimanjaro, Mweka route, 27 July 1968, *Bigger* 2039!; Lushoto District: Shume-Magambe Forest Reserve, 2 May 1987, *Kisena* 577!; Njombe District: forests around Milo village, 13 Nov. 1987, *Mwasumbi et al.* 13691!

DISTR. **U** 2, 3; **K** 4, 5, 7; **T** 1–3, 6, 7; Burundi, Cameroon, Congo-Kinshasa, Ethiopia, Sudan, Zambia, Malawi, Zimbabwe

HAB. Wet montane forests, swamp forests, bamboo zone and giant heath zone; (1150–)1850–2700(–3500) m

USES. None recorded for our area

CONSERVATION NOTES. Widespread; least concern (LC)

SYN. *Aspidium kilemense* Kuhn, Fil. Deck.: 24 (1867), as '*kilmense*'
Nephrodium kilemense (Kuhn) Baker in Hook. & Baker, Syn. Fil.: 498 (1874), as '*kilmense*'
Nephrodium lastii Baker in Ann. Bot. 5: 324 (1891); Sim, Ferns S. Afr., ed. 2: 109 (1915). Type: Mozambique, Namuli, Makua Country, 1887, *Last* s.n. (K!, holo., 2 sheets)
Aspidium lastii (Baker) Hieron. in P.O.A. C: 85 (1895)
Dryopteris lastii (Baker) C.Chr., Ind. filic.: 274 (1905)
Dryopteris platylepis Rosenst. in F.R. 4: 4, 5 (1907). Type: Tanzania, Kilimanjaro, *Daubenberger* 37 (M!, holo.; M!, iso.)

NOTE. Vida, in Helv. Chim. Acta 56: 2129 (1973) report the species as diploid: 2n = ± 82.

6. **Dryopteris lewalleana** *Pic.Serm.* in B.J.B.B. 55: 158, fig. 3 (1985). Type: Burundi, Bujumbura Province, Mwisare road, *Lewalle* 5482 (FT-Pic.Serm., holo.; BR!, FT-Pic.Serm., iso.)

Terrestrial or epilithic; rhizome short-decumbent, sparsely branched, up to 21 mm in diameter, with stramineous to ferrugineous ovate to linear scales up to 18 × 4 mm, apex caudate to acuminate, entire or with a few hairs and pyriform glands. Fronds tufted, suberect to arching, up to 1.4 m long; stipe proximally castaneous, stramineous to greenish higher up, up to 75 cm long and 10 mm in diameter, with dense to moderate stramineous to ferrugineous scales, ovate to filiform, to 25 × 6 mm, often with a few multicellular hairs and glands; lamina herbaceous, ovate to deltate, 2- to 3-pinnate, up to 64 cm long; rachis stramineous, becoming narrowly winged towards apex, eglandular or sparsely glandular, and moderately to sparsely set with scales to 8 × 1.5 mm and hairs; pinnae in up to 14 stalked pairs, basal pinna pair inaequilaterally triangular to broadly ovate or deltate, up to 32 × 25 cm, basal pair shorter or longer than pair above, with up to 10 stalked pinnule pairs; pinna-rachis narrowly winged for most of length, eglandular or sparsely glandular, abaxially with hairs and scales similar to, but smaller than those on rachis; pinnules narrowly triangular to lanceolate, distally basiscopically decurrent, acroscopic pinnule on basal pinnae up to 94 × 34 mm, basiscopic pinnule on basal pinnae up to 145 × 52 mm, with up to 2 pairs of stalked segments; costa glandular or eglandular, narrowly winged, variously set with isocytic or moniliform hairs and scales; segments spaced, inaequilaterally narrowly ovate-obtuse to oblong-obtuse, basiscopically decurrent, up to 28 × 9 mm, shallowly to deeply lobed, lobes serrate, adaxially glabrous, with a few glands along and between veins and/or with a few filiform scales along costa, abaxially with hairs and scales to 3 × 1.2 mm, and/or oblong to clavate glands. Sori medial on predominantly anadromous vein branches, essentially 2-seriate on segments, discrete, circular, up to 1.6 mm in diameter; sporangium stalk simple, with one or more glandular cells, or with a long uniseriate hair; indusium persistent, brown, firmly herbaceous, reniform, glabrous, margin repand, rarely lacerate, often strongly revolute, up to 1.6 mm in diameter. Spores ellipsoidal, monolete, perispore folded into low tubercules or reticulate ridges, finely rugose to ruminate, 28–50 × 20–31 mm.

UGANDA. Kigezi District: Kachwekano farm, Sept. 1949, *Purseglove* 3112! & Rubaya forest, *A.S. Thomas* 4254!

KENYA. Northern Frontier District: summit of Kulal, 25 July 1958, *Verdcourt* 2243!; Kiambu District: SE Aberdares forest, Kitikuyu, *Gardner* 1269!; Machakos/Kitui District: Ukambani, Dec. 1893, *Scott-Elliot* 6524!

TANZANIA. Arusha District: Mt Meru, Engarenanyuki River, 27 Dec. 1966, *Vesey-Fitzgerald* 5042!; Ufipa District: Namwele, *Bullock* 2567!; Njombe District: Nyehamutwe, Lunganya [Hagafilo] village, 8 Nov. 1987, *Mwasumbi et al.* 13532!;

DISTR. **U** 2; **K** 1, 4; **T** 2–4, 7, 8; Congo-Kinshasa, Burundi, Ethiopia, Zambia, Malawi, Mozambique, Zimbabwe, Swaziland, South Africa

HAB. Stream and riverbanks, riverine forests, hillside thickets, *Terminalia*- and *Brachystegia*-*Uapaca*-woodland, as well as montane forests; (800–)1400–1850(–2300) m

USES. None recorded for our area

CONSERVATION NOTES. Widespread; least concern (LC)

SYN. *Lastrea pentagona* T.Moore in J. Bot. 5: 227 (1853). Type as for *D. lewalleana*
Aspidium pentagonum (T.Moore) Kuhn, Fil. Afr.: 139 (1868)
Dryopteris inaequalis (Schltdl.) Kuntze var. *atropaleacea* Schelpe in Bol. Soc. Brot., Sér. 2, 4: 213 (1967). Type: Tanganyika, Ufipa District: Mbisi forest, *Vesey-Fitzgerald* 1390 (BOL!, holo.)
[*Dryopteris inaequalis* sensu Schelpe, F.Z., Pterid.: 221 (1970), *non* (Schltdl.) Kuntze]

NOTE. The new name *D. lewalleana* had to be chosen as *D. pentagona* is already occupied by *Dryopteris pentagona* Bonap., Notes pterid. 5: 64, 65 (1917).

7. **Dryopteris manniana** (*Hook.*) *C.Chr.,* Ind. filic.: 276 (1905); Tardieu in Mém. I.F.A.N. 28: 149, t. 29/4–5 (1950); Alston, Ferns W.T.A.: 70 (1959); Schelpe, F.Z., Pterid.: 223, t. 63 (1970); Schelpe & Diniz, Fl. Moçamb., Pterid.: 240 (1979); Jacobsen, Ferns S. Afr.: 440, t. 331, map 161 (1983); Pichi Sermolli in B.J.B.B. 55: 163 (1985); Burrows, S. Afr. Ferns: 303, t. 50/6, fig. 70/307, map (1990); Faden in U.K.W.F.: 36 (1994); J.P. Roux, Conspect. southern Afr. Pterid.: 125 (2001). Type: Bioko [Fernando Po], on the peak, 2000 ft, 1860, *Mann* s.n. (K!, holo.)

Terrestrial; rhizome erect to short-decumbent, mostly unbranched, to 8 mm in diameter, with closely spaced stipe bases and brown to ferrugineous subulate scales to 22 × 3 mm, closely set with short teeth. Fronds tufted, 4–7 per plant, arching, to 1 m long; stipe greenish to stramineous, up to 48 cm long and up to 5 mm in diameter, proximally densely scaly, higher up moderately scaly, scales stramineous to ferrugineous, up to 15 × 4.5 mm, basally frequently with filiform outgrowths, often with scattered glandular cells, denticulate; lamina herbaceous, ovate, up to 53 cm long, to 2-pinnate-pinnatifid, proliferous, generally with one or more scaly buds adaxially along rachis near lamina apex; rachis greenish to stramineous, narrowly winged towards apex, with ferrugineous to stramineous filiform to lanceolate scales up to 7 × 1.8 mm, denticulate, basally with filiform outgrowths, higher up closely set with short teeth; pinnae in up to 12 stalked pairs, basal pinna pair mostly longest, inaequilaterally ovate, narrowly ovate to oblong-acuminate towards lamina apex, basiscopically developed, up to 250 × 95 mm, with up to 3 stalked pinnule pairs; pinna-rachis narrowly winged for most of length, abaxially with sparse to moderate stramineous to ferrugineous filiform to narrowly oblong scales up to 5 × 1 mm; pinnules symmetric or inaequilaterally narrowly ovate to ovate, basiscopically decurrent, pinnatifid, often basiscopically developed, acroscopic pinnule on basal pinnae up to 50 × 17 mm, basiscopic pinnule on basal pinnae up to 56 × 23 mm, lobes broadly oblong-obtuse, shallowly lobed, dentate, glabrous adaxially, abaxially sparsely set with hairs and scales up to 2 mm long along veins. Sori inframedial on vein branches, discrete at maturity, circular, to 1.5 mm in diameter; sporangium stalk simple or haired; exindusiate. Spores with short and long ridges, ruminate, 40–58 × 26–32 mm. Fig. 13: 3.

UGANDA. Ankole District: Igara County, sawmill W of Rubuzigye in Kalinzu forest, 19 Sept. 1969, *Faden* 69/1171!; Bushenyi, Kasyoha-Kitomi Forest Reserve, 16 Nov. 1994, *Poulsen* 717!; Kigezi District: Bwindi National Park, North Sector (Kayonza), near Ishasha River, 22 March 1995, *Poulsen* 806!

KENYA. Meru District: Nyambeni Hills, 1.5 km N of Maua on Maua–Kangeta road, 31 May 1969, *Faden et. al.* 69/650!; Kericho District: Itari River, Sept. 1949, *H.D. van Someren* 291!; Teita District: Mbololo Hill, Mraru Ridge, 5 July 1969, *Faden et al.* 69/8241!

TANZANIA. Lushoto District: Shume-Magamba Forest Reserve, 2 May 1987, *Kisena* 620!; Kilosa District: Ukaguru Mts, Mamiwa Forest Reserve, 1 km N of Mandege Forest Station, 8 Aug. 1972, *Mabberley* 1415!; Iringa District: Mwanihana Forest Reserve above Sanje village, 10 Oct. 1984, *D.W. Thomas* 3802!

DISTR. **U** 2; **K** 4, 5, 7; **T** 3, 5–7; Sierra Leone, Liberia, Ivory Coast, Nigeria, Bioko, Cameroon, Malawi, Mozambique and Zimbabwe; Madagascar

HAB. Terrestrial in moist forests and riparian forests; (1000–)1450–1850(–2250) m

USES. None recorded for our area

CONSERVATION NOTES. Widespread; least concern (LC)

SYN. *Polypodium mannianum* Hook., Sp. Fil. 4: 253 (1862)
Phegopteris manniana (Hook.) Kuhn, Fil. Afr.: 123 (1868)

NOTE. Vida, in Helv. Chim. Acta 56: 2130 (1973) reports the species as tetraploid: 2n = ± 164.

8. **Dryopteris pentheri** (*Krasser*) *C.Chr.*, Ind. filic.: 284 (1905); Faden in U.K.W.F. ed. 2: 36 (1994). Type: South Africa, Van Reenen's Pass, 4 March 1895, *Krook s.n.*, sub Penther Plantae Austro-Africanae 36 (W 8042!, lecto.; BM!, isolecto.) designated by Pichi Sermolli (166: 1985)

Terrestrial or epilithic; rhizome short-decumbent, sparsely branched, up to 20 mm in diameter, with closely spaced stipe bases and with stramineous to ferrugineous linear, oblong or narrowly ovate scales to 37 × 6 mm, with long twisted filiform outgrowths. Fronds tufted, erect to arching, up to 1.8 m long; stipe proximally castaneous, brown to stramineous higher up, up to 49 cm long and 10 mm in diameter, near base with dense stramineous to ferrugineous narrowly lanceolate to filiform scales up to 40 × 7 mm, with long twisted filiform outgrowths and often also with scattered glands, scales higher up fugaceous; lamina herbaceous, ovate to ovate-triangular, up to 78 cm long, 2-pinnate to 3-pinnate; rachis stramineous, becoming narrowly winged towards apex, with fugaceous narrowly lanceolate to linear scales up to 7 × 2 mm; pinnae in up to 16 stalked pairs, basal pinna pair inaequilaterally ovate to oblong-acuminate towards lamina apex, basal pair longest, mostly basiscopically developed, up to 315 × 185 mm, with up to 9 stalked pinnule pairs; pinna-rachis narrowly winged distally, abaxially with scales up to 5 mm × 1.6 mm and hairs of three types (unicellular, bicellular and pluricellular simple or branched); pinnules lanceolate to oblong-acuminate, basiscopically decurrent, acroscopic pinnule on basal pinnae up to 90 × 35 mm, basiscopic pinnule on basal pinnae up to 112 × 42 mm; segments ovate to oblong-obtuse, basiscopically decurrent, up to 22 × 9 mm, lobed, lobes serrate, adaxially glabrous, with oblong glands along and between veins, or with a few hairs along costule, abaxially sparsely set with glands and hairs mostly along costule and veins. Sori medial on predominantly anadromous vein branches, discrete, circular, up to 1.8 mm in diameter at maturity, essentially uniseriate; sporangium stalk simple, or with one or more glandular cells, but mostly with a long multicellular, uniseriate hair; indusium persistent, pale brown, firmly herbaceous, reniform, entire, repand, or erose, (rarely glandular along margin), often strongly revolute, up to 1.8 mm in diameter. Spores with perispore folded into tubercles or reticulate ridges, finely rugose to ruminate, 38–60 × 27–40 mm.

UGANDA. Ruwenzori, Mihunga, no date, *Loveridge* 358!; Ankole District: Buhweju, Nyagoma-Rugongo, 9 Feb. 1990, *Rwaburindore* 2949!; Kigezi District: Kigezi, Kirata gap, *Chandler & Hancock* 2532!

KENYA. Trans-Nzoia District: Cherangani, Kapolet, Aug. 1963, *Tweedie* 2681!; Machakos District: Mt Nzaui, 16 Feb. 1969, *Faden & Evans* 69/191!; Teita District: Taita Hills, road from Weruga to Mgange, 19 Nov. 1969, *Bally et al.* 2972!

TANZANIA. Mbulu District: Mt Hanang, below Werther's Peak, 12 Feb. 1946, *Greenway* 7726!; Buha District: Kasakela Reserve, 20 Nov. 1962, *Verdcourt* 3387!; Lindi District: Rondo Forest Reserve, 10 Feb. 1991, *Bidgood et al.* 1456!

DISTR. **U** 2; **K** 3–5, 7; **T** 2–4, 6–8, Guinea-Bissau, Bioko, Cameroon, Burundi, Sudan, Ethiopia, Zambia, Malawi, Mozambique, Zimbabwe, Swaziland, Lesotho and South Africa; Madagascar

HAB. Terrestrial or epilithic, in woodland, riverine and montane forests, roadcuttings, but at higher elevations (> 2000 m) in grasslands, moorlands and giant heath zone; (700–)900–2700(–2900) m

USES. None recorded for our area

CONSERVATION NOTES. Widespread; least concern (LC)

SYN. *Nephrodium pentheri* Krasser in Ann. K. K. Nat. Hofmus. 15: 5 (1900)
[*Dryopteris inaequalis* sensu Schelpe, F.Z., Pterid.: 221 (1970), *non* (Schltdl.) Kuntze].

NOTE. Vida, in Helv. Chim. Acta 56: 2129 (1973), under *D. inaequalis*, reports the species as tetraploid: $2n = 164 \pm 4$ and $2n = \pm 164$.

9. **Dryopteris rodolfii** *J.P.Roux* in *Webbia* 59: 143, figs. 3 & 4 (2004). Type: Ethiopia, Bale mountains, above Goba, *G. & S. Miehe* 3239 (K!, holo.)

Terrestrial; rhizome not seen but probably short and decumbent. Frond up to 76 cm long; stipe proximally castaneous, stramineous higher up, up to 31 cm long and 5 mm in diameter, proximally with dense ferrugineous ovate scales to 19 × 5 mm, caudate, denticulate and with a few filiform outgrowths, less dense distally; lamina firmly herbaceous, broadly ovate to deltate-acuminate, up to 46 × 38 cm, to 2-pinnate-pinnatifid; rachis stramineous, sparsely to moderately scaly with spreading ferrugineous linear to ovate scales up to 9 × 2.5 mm, caudate, denticulate, often with a few glandular cells, and with a few filiform outgrowths; pinnae in up to 8 petiolate pairs, basal pair longest or slightly shorter than next pair higher up, basal pair basiscopically developed, inaequilaterally triangular to narrowly ovate, linear-acuminate towards apex, up to 217 × 80 mm, with up to 3 stalked pinnule pairs; pinna-rachis stramineous, narrowly winged for most of length, wing with a conspicuously thickened margin, abaxially with scales up to 4 × 1.2 mm; pinnules eventually adnate and basiscopically decurrent towards apex, lanceolate to oblong-obtuse, pinnatifid but mostly lobed, up to 60 × 20 mm, adaxially ± glabrous, abaxially with ferrugineous to stramineous scales and hairs, scales up to 2.4 × 0.8 mm, confined to costa and veins, hairs up to 1.3 mm long; lobes oblong-obtuse to oblong-truncate, dentate, 6–10 mm long, 3–5 mm wide. Sori inframedial, discrete, but often touching at maturity in smaller laminae, essentially 2-seriate on pinnules, but often 2-seriate on basal lobes in larger and more divided laminae, circular, up to 1.8 mm in diameter at maturity; sporangium stalk simple or with a uniseriate, pluricellular hair; indusium brown to ferrugineous, persistent, broadly ovate to reniform, basally entire to repand, erose distally, 1.6–2.6 mm in diameter. Spores dark brown, regularly tuberculate, exospore 42–56 × 24–38 µm.

KENYA. Mt Kenya National Park, near Sirimon gate, 24 Aug. 1986, *Jermy* 17507! & Mt Kenya, Forest End camp, 29 Aug. 1942, *McLoughlin* 696!
DISTR. **K** 4; Ethiopia
HAB. *Hagenia*-bamboo zone; 2750–2950 m
USES. None recorded for our area
CONSERVATION NOTES. Distribution restricted, status unknown.

10. **Dryopteris ruwenzoriensis** *Fraser-Jenkins* in Bull. Brit. Nat. Hist. Mus., Bot. 14: 204, fig. 4 (1986). Type: Congo-Kinshasa, Ruwenzori, *Humbert* 8825 (BM!, holo.)

Terrestrial or epilithic; rhizome short-decumbent to suberect, up to 15 mm in diameter, with crowded persistent stipe bases, and castaneous to ferrugineous narrowly ovate to lanceolate scales, up to 12 × 4 mm, often with a few scattered glandular cells. Fronds tufted, erect to arching, up to 1.4 m long; stipe proximally castaneous, stramineous higher up, aerophores in larger plants often displayed as paler dorso-lateral lines, up to 72 cm long and 9 mm in diameter, proximally with dense narrowly ovate to lanceolate scales up to 12 × 2.5 mm, denticulate, moderately scaly higher up and variously set with hairs and fugaceous minute membranous scales; lamina firmly herbaceous, ovate to triangular, up to 640 × 540 mm, 2-pinnate-pinnatifid to 3-pinnate; rachis stramineous, narrowly winged near apex, with fugaceous hairs and scales up to 5 × 1 mm; pinnae in up to 5 stalked pairs, basal pair conspicuously basiscopically developed, inaequilaterally narrowly ovate to triangular, basal pair longest, up to 310 × 125 mm, with up to 2 stalked pinnule pairs; pinna-rachis narrowly winged distally, abaxially with stramineous scales; pinnules progressively more broadly attached and basiscopically decurrent towards apex, lanceolate to oblong-acuminate, pinnatifid to deeply lobed, acroscopic pinnule on basal pinna up to 51 × 20 mm, basiscopic pinnule on basal pinna up to 88 × 26 mm; costa narrowly winged along entire length, scales up to 1.5 × 0.8 mm; segments closely spaced but not imbricate, oblong-obtuse to oblong-truncate, up to 16 × 9 mm, shallowly and inconspicuously lobed, lobe apices conspicuously and acutely dentate, adaxially glabrous or with a few hairs and hair-like scales along costule, abaxially variously set with subulate scales and hairs to 0.7 mm long along veins. Sori circular, medial to inframedial, predominantly on anadromous vein branches, discrete, up to 1.5 mm in diameter at maturity, essentially 2-seriate on segments; sporangium stalk simple or haired; exindusiate. Spores without ridges, echinate, 36–54 × 22–34 µm. Fig. 14.

UGANDA. Ruwenzori Mountains, Mobuku Valley, no date, *Esterhuysen* 25178a! & Bujuku Valley, near Bigo camp, 25 March 1948, *Hedberg* 471!
DISTR. **U** 2; Rwanda and Congo-Kinshasa; endemic to the Ruwenzori
HAB. Moist forest with *Podocarpus*, bamboo and giant heath zone; 1800–3550 m
USES. None recorded for our area
CONSERVATION NOTES. Distribution restricted, least concern (LC)

11. **Dryopteris schimperiana** (*A.Br.*) *C.Chr.*, Ind. Filic.: 91, 291 (1905); Faden in U.K.W.F. ed. 2: 36 (1994). Type: Ethiopia, Scholoda Mts, *Schimper* 6 (B?, holo.; BM!, H, K! (2 sheets), M! (2 sheets), iso.)

Plants terrestrial; rhizome short-decumbent, up to 18 mm in diameter, with crowded stipe bases and scales up to 35 × 5 mm, brown to ferrugineous, linear to narrowly lanceolate, with scattered oblong glands, and irregularly set with long twisted filiform outgrowths. Fronds tufted, 5–7 per plant, erect to arching, up to 1.6 m long; stipe proximally castaneous, stramineous to greenish higher up, up to

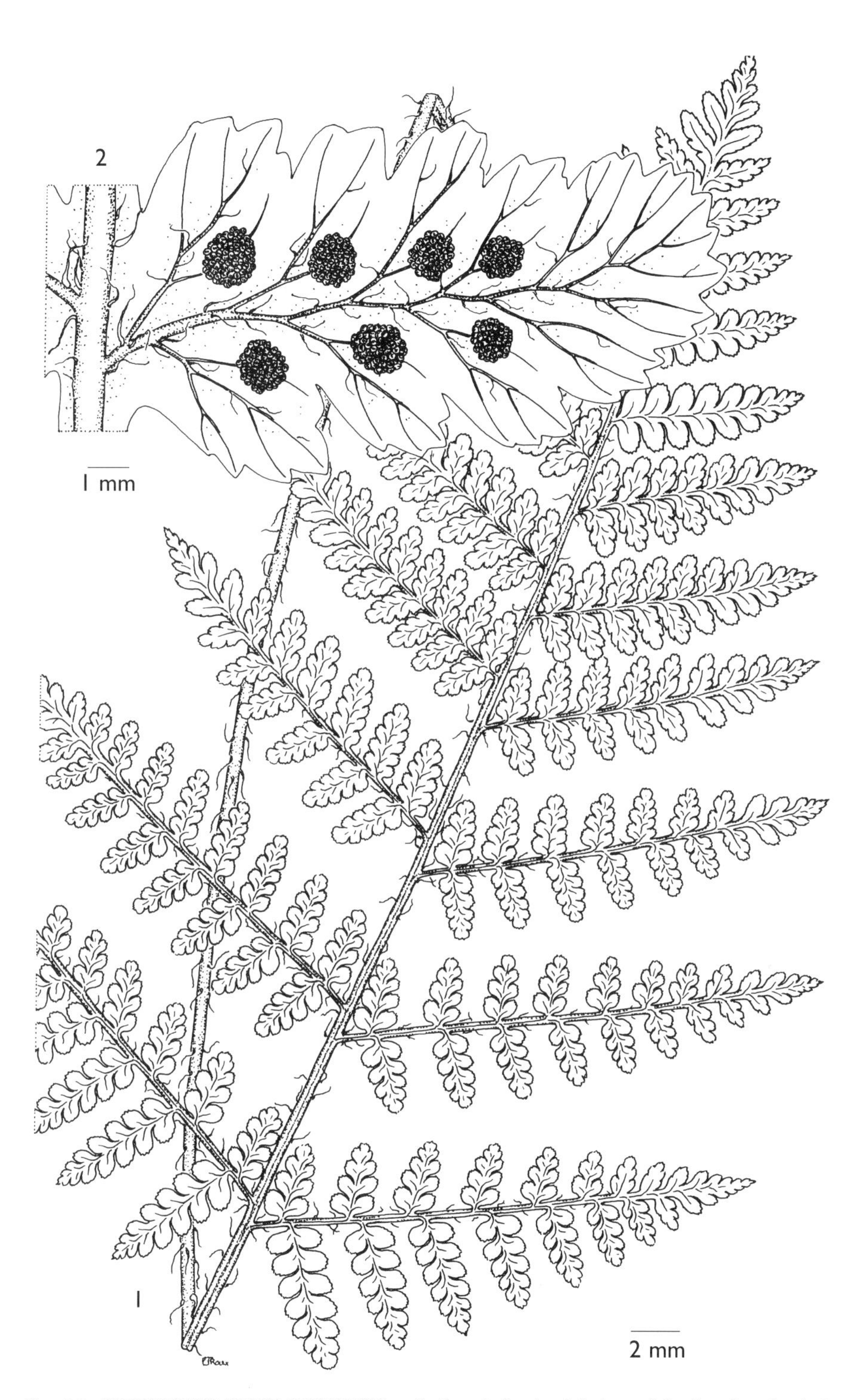

FIG. 14. *DRYOPTERIS RUWENZORIENSIS* — **1**, frond; **2**, abaxial view of fertile pinnule. 1–2 from *Osmaston* 1687. Drawn by J.P. Roux.

87 cm long and 8 mm in diameter, proximally densely scaly, scales higher up fugaceous, larger scales up to 22 × 3 mm, narrowly lanceolate to filiform, irregularly set with long, twisted, pluricellular filiform outgrowths, often also with oblong glands; lamina firmly herbaceous, ovate to broadly ovate, up to 73 × 45 cm, up to 3-pinnate; rachis stramineous, with scales up to 8 × 1.2 mm, glabrescent; pinnae in up to 14 stalked pairs, basal pair mostly basiscopically developed, inaequilaterally ovate to narrowly ovate, narrowly ovate to oblong-acuminate towards lamina apex, basal pair longest, up to 23 × 14.5 cm, with up to 10 stalked pinnule pairs; pinna-rachis narrowly winged distally, abaxially initially with scales up to 6 × 1 mm; pinnules oblong-acuminate to narrowly trullate, obtuse, lobed, to 1-pinnate, acroscopic pinnule on basal pinnae up to 72 × 19 mm, basiscopic pinnule on basal pinnae up to 84 × 22 mm, rarely with up to 2 stalked segment pairs; costa adaxially shallowly sulcate, narrowly winged, variously set with scales and isocytic hairs; segments and lobes inaequilaterally oblong-obtuse to oblong-acute, basiscopically decurrent, up to 10 × 5 mm, shallowly obtusely dentate or serrate, adaxially glabrous or with few hairs or scales, abaxially sparsely to moderately set with scales and hairs, scales similar, but smaller than those on pinna-rachis, hairs up to 1 mm long. Sori circular, inframedial, discrete, or the sporangia often touching at maturity, up to 2 mm in diameter; sporangium stalk simple, with a single glandular cell, or haired; indusium persistent, brown to ferrugineous, reniform and strongly revolute, entire or glandular along margin, up to 2 mm in diameter. Spores with low reticulate ridges and bulges, rugose to ruminate, (32–)41(–52) × (22–)28(–34) mm.

UGANDA. Toro District: Mt Ruwenzori, Mihunga, 13 Jan. 1939, *Loveridge* 354!; Kigezi District: Kisaba Gap, Dec. 1938, *Chandler & Hancock* 2539! & Kanaba Pass, 28 Dec. 1959, *Lind & McLeay* 3047!

KENYA. Trans-Nzoia District: Cherangani, Kabolet river, June 1964, *Tweedie* 2844! & Marun River, upper waters in Cherangani, March 1965, *Tweedie* 3012!

TANZANIA. Kilimanjaro, forest above Mandara Hut, 10 Feb. 1994, *Grimshaw* 94230!; Kilosa District: Ukaguru Mts, 31 July 1972, *Mabberley* 1319!; Mbeya District: Poroto Mts, Kikondo camp, 20 Jan. 1961, *Richards* 13968!

DISTR. **U** 2; **K** 3; **T** 2, 6, 7; Congo-Kinshasa, Rwanda, Burundi, Sudan, Eritrea, Ethiopia and Malawi

HAB. Evergreen montane and riverine forest, bamboo, Hagenia and giant heath zone; 1400–3000 m

USES. None recorded for our area

CONSERVATION NOTES. Widespread; least concern (LC)

SYN. *Aspidium schimperianum* A.Br. in Flora 45: 708 (1841)
Polystichum schimperianum (A.Br.) Keyserl., Polyp. Herb. Bunge.: 44 (1873)

NOTES. Vida, in Helv. Chim. Acta 56: 2130 (1973) report the species as diploid: 2n = ± 82.

12. **Dryopteris tricellularis** *J.P.Roux* in K.B. 57: 735, fig. 1 & 2 (2002). Type: Uganda, Mt Elgon, *Dummer* 3557 (NBG!, holo.; BOL!, K!, iso.)

Terrestrial; rhizome short-decumbent, up to 15 mm in diameter, with closely spaced stipe bases and dark brown ovate to lanceolate scales up to 15 × 7 mm, with short and long filiform outgrowths, and/or with scattered unicellular glands. Fronds tufted, suberect to arching, up to 1.1 m long; stipe proximally castaneous, ferrugineous to stramineous higher up, up to 57 cm long and 9 mm in diameter, proximally with dense brown to dark broadly ovate to lanceolate scales up to 15 × 7 mm, with a few scattered unicellular glands, or with uniseriate filiform outgrowths, apex flagelliform, higher up moderately to sparsely set with stramineous to ferrugineous scales and hairs; lamina firmly herbaceous, ovate to narrowly triangular, up to 55 × 44 cm, to 3-pinnate; rachis stramineous, narrowly winged towards apex, initially moderately to densely set with scales and hairs, scales stramineous, narrowly lanceolate to filiform, up to 6 × 1.8 mm; pinnae in up to 18 stalked pairs, basal pair inaequilaterally ovate to oblong-acuminate, lanceolate to oblong-acuminate towards lamina apex, basal pair slightly shorter or longer than pair above, basiscopically developed in lower half of lamina, up to 330 × 180 mm, with up to 11 stalked pinnule pairs; pinna-rachis narrowly winged along entire length in smaller specimens, abaxially moderately to densely set with stramineous to ferrugineous ovate to filiform scales up to 4 × 1.2 mm and hairs; pinnules narrowly triangular to oblong-obtuse, pinnatifid to 1-pinnate, acroscopic pinnule on basal pinnae up to 80 × 35 mm, basiscopic pinnule on basal pinnae up to 90 × 28 mm; costa narrowly winged, abaxially moderately to densely set with scales similar, but smaller than those on pinna-rachis, and with hairs; segments spaced, oblong-obtuse, basiscopically decurrent, up to 12 × 4 mm, lobed, lobes serrate, adaxially glabrous or sparsely set with hairs, abaxially closely set with hairs along costae, costules, and veins. Sori medial to inframedial, discrete at maturity, circular, up to 1.2 mm in diameter; sporangium stalk simple, glandular, or haired; exindusiate or indusium brown, cordate to reniform, repand to irregularly lobed, up to 1 mm in diameter. Spores with prominent reticulate ridges, with granulate deposits, 32–52 × 22–38 mm.

UGANDA. Toro District: Bwamba Pass, 16 Nov. 1935, *A.S. Thomas* 1442!; Mt. Elgon, Jan. 1918, *Dummer* 3352! & 3557!

KENYA. Mt Elgon, crater at Mayi ya Moto, 15 May 1948, *Hedberg* 907!; Mt Kenya, above Timau, 28 Aug. 1942, *McLoughlin* 683! & Kamweti track, crossing of Gathiba River, 26 Jan. 1969, *Faden* 69/112!

TANZANIA. Morogoro District: Nguru Mountains, between Kombola and Maskah Mafulumla mountains, 20 Aug. 1971, *Schlieben* 12244!; Rungwe District: Rungwe forest, June 1957, *Watermeyer* 37!

DISTR. **U** 2, 3; **K** 3, 4; **T** 6, 7; appears to be restricted to the mountainous regions of tropical East Africa and the Ahmar mountains in Ethiopia

HAB. Montane and riverine forests with *Podocarpus*, bamboo zone; 1700–3600 m

USES. None recorded for our area

CONSERVATION NOTES. Widespread; least concern (LC)

SYN. [*Dryopteris inaequalis* sensu Hedberg, A.V.P.: 26 (1957), *non* (Schltdl.) Kuntze]

12. **ARACHNIODES**

Blume, Enum. Pl. Jav. 2: 241 (1828)

Rhizome suberect to creeping. Fronds tufted or spaced; stipe with brown or ferrugineous scales basally; lamina broadly triangular to pentagonal, herbaceous to coriaceous; lower pinnae much developed basiscopically (in African species), much dissected, the ultimate segments dentate-aristate (in African species); veins free; rachis with ridges on the ventral surface not continuous with leaf margin. Sori circular with rounded reniform indusia.

A genus of about 200 described species (but probably far fewer in reality) mostly in the Himalayas and China, others in Australia and America, one in Africa and also one in Madagascar (see note).

Arachniodes webbiana (*A.Braun*) *Schelpe* in Bol. Soc. Brot. sér. 2, 41: 203 (1967); J.P. Roux, Consp. S. Afr. Pterid.: 127 (2001). Type: Madeira, ad Sanctum Annam et Sanctum Vincentem, *sine coll.* (B90924, holo.) (fide J.P. Roux)

Rhizome creeping, 60–90 cm long, up to 7 mm in diameter, set with brown, linear-attenuate entire rhizome-scales up to 6 mm long. Fronds spaced, arching, firmly herbaceous, 35–90 cm tall; stipe stramineous, densely scaly basally, up to 50 cm long; lamina broadly ovate-triangular, up to 45 × 30 cm, acuminate, 4-pinnatifid basally, 2-pinnatifid above, the basal pinnae largest and much developed basiscopically; rachis straw-coloured, set with scattered scales; upper pinnae narrowly lanceolate-attenuate; basal pinnae unequally and broadly ovate-triangular, up to 19 cm broad;

FIG. 15. *ARACHNIODES WEBBIANA* subsp. *FOLIOSA* — **1**, habit, × ½; **2**, pinna, × 2. 1–2 from *Chase* 7818. Drawn by Monika Shaffer-Fehre.

ultimate pinnatifid segments narrowly rhombic and strongly aristate-dentate, glabrous except for hair-pointed ovate-lanceolate scales along costae and costules, and smaller hair-like scales scattered along veins. Sori up to 1 mm in diameter; indusium entire, minutely papillose, but in some specimens very deciduous and whole fronds may reveal very few or even none at certain stages. Fig. 15.

SYN. *Aspidium webbianum* A. Braun in Flora 24: 711 (1841)

subsp. **foliosa** (*C.Chr*) *Gibby, Rasbach, Reichstein, Widén & Viane* in Bot. Helvet. 102: 243 (1992); J.P. Roux, Consp. S. Afr. Pterid.: 127 (2001). Type: Kenya, Aberdare Mts, Kinangop, *Alluaud* 255 (BM, holo!)

Virtually identical in macromorphological characters but tetraploid not diploid and with small but distinct differences in microcharacters including epidermal cells and the scales on undersides of pinnules (see Gibby et al. for details).

UGANDA. Kigezi District: NW face of Muhavura, Jan. 1933, *Eggeling* 1049!; Mt Elgon, Sasa Trail, 28 Dec. 1996, *Wesche* 481! & W Elgon, Bufumbo, Dec. 1950, *H.D. van Someren* 417!

KENYA. Naivasha District: Aberdares, S Kinangop, 24 Nov. 1957, *Molesworth-Allen* 3636!, 3635!; Kiambu District: Limuru, Feb. 1915, *Dummer* 1726!; Teita District: Mt Kasigau, path from Rukanga up mountain, 5 Apr. 1969, *Faden et al.* 69/472!

TANGANYIKA. Moshi District: S Kilimanjaro, Una Stream, 22 Jan. 1934, *Schlieben* 4619!; Pare District: Gonja, Shengena Forest Reserve, 29 Feb. 1988, *Kisena* 339!; Lushoto District: W Usambaras, 14 Sept. 1981, *Mtui & Sigara* 59!; Morogoro District: W slope of Nguru Mts, above Maskati, 17 Mar. 1988, *Bidgood et al.* 538!

DISTR. **U** 2, 3; **K** 1 (Mt Nyiro), 3–7; **T** 2, 3, 6; Rwanda, Malawi, Zimbabwe, South Africa (E Cape Province to Transkei, Natal & Transvaal)

HAB. Montane and intermediate forest with *Podocarpus*, bamboo, etc., stream gulleys, high rainfall areas; (?1250–)1380–2600(–3130 Elgon) m

USES. None recorded for our area

CONSERVATION NOTES. Widespread; least concern (LC)

SYN. [*Polystichum aristatum* sensu Sim, Ferns S. Afr. ed. 2: 119, t. 30 (1915), non (Forst.) Presl]
Dryopteris foliosa C.Chr. in Dansk. Bot. Ark. 9 (3): 61 (1937)
Arachniodes foliosa (C.Chr.) Schelpe in Bol. Soc. Brot. sér. 2, 41: 203 (1967); Schelpe, F.Z., Pterid.: 228, t. 65 (1970); Faden, in U.K.W.F.: 48 (1974); Kornaś, Distr. Ecol. Pterid. Zambia: 106, fig. 70B (1979); W. Jacobsen, Ferns S. Afr.: 450, t. 340, map 167 (1983); Pic. Serm. in B.J.B.B. 55: 175 (1985);' Schelpe & N.C. Anthony, F.S.A., Pterid.: 259, fig. 89, map 227 (1986); J.E. Burrows, S. Afr. Ferns: 318, t. 53/3, fig. 323, map (1990); Faden in U.K.W.F. ed. 2: 35 (1994). Type: Kenya, Aberdare Mts, Kinangop, *Alluaud* 255 (BM, holo.!)

NOTE. A photograph of the type of *Rumohra humbertii* Tardieu described from Madagascar looks very similar to *A. webbiana* subsp. *foliosa*. *A. webbiana* subsp *webbiana* is confined to Madeira.

INDEX TO DRYOPTERIDACEAE

New names validated in this part

Tectaria torrisiana *Shaffer-Fehre* **sp. nov.**

GEOGRAPHICAL DIVISIONS OF THE FLORA

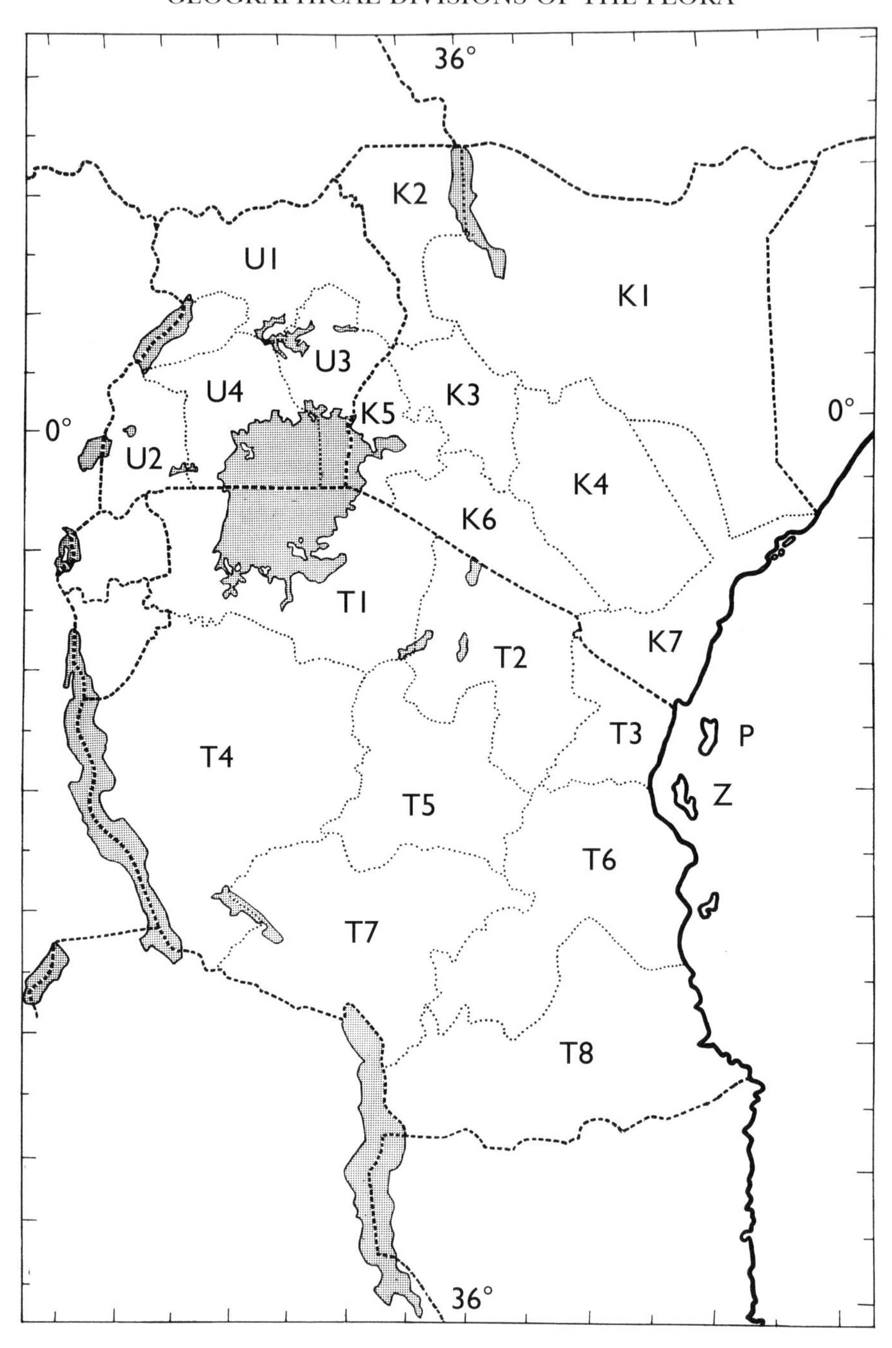

First published in 2007 by
Royal Botanic Gardens, Kew
Richmond, Surrey, TW9 3AB, UK
www.kew.org

ISBN 978 1 84246 188 4

British Library Cataloguing in Publication Data
A catalogue record for this book is available from the British Library

Design and typesetting by Margaret Newman,
Kew Publishing, Royal Botanic Gardens, Kew.

Printed in the UK by Hobbs the Printers

For information or to purchase all Kew titles please visit
www.kewbooks.com or email publishing@kew.org

All proceeds go to support Kew's work in saving the world's plants for life